KATHERINE PLAKE HOUGH

MICHAEL ZAKIAN

Transforming
the WESTERN IMAGE

in 20th Century American Art ■

Palm Springs Desert Museum

TRANSFORMING THE WESTERN IMAGE
IN 20TH CENTURY AMERICAN ART

Palm Springs Desert Museum
Palm Springs, California
February 21–April 26, 1992

Boise Art Museum
Boise, Idaho
May 23–July 12, 1992

Tucson Museum of Art
Tucson, Arizona
August 7–October 4, 1992

The Rockwell Museum
Corning, New York
October 25–December 20, 1992

This project is funded in part
by the Palm Springs Desert Museum's
Contemporary Art Council

Published by the
Palm Springs Desert Museum
101 Museum Drive
Palm Springs, California 92262

Distributed by
University of Washington Press
P.O. Box 50096
Seattle, Washington 98145-5096

FRONT COVER
Top, left to right: Arthur Dow, *Bright Angel Canyon;* Wayne
Thiebaud, *Half Dome Evening;* Georgia O'Keeffe, *The Red Hills
Beyond Abiquiu.*
Bottom, left to right: Roy Lichtenstein, *American Indian
Theme III;* Red Grooms, *Great Western Act;* William Allan,
Shadow Repair for the Western Man.

BACK COVER
Top, left to right: Conrad Buff, *Jawbone Canyon;* Mark Tobey,
Mexican Ritual; Elliott Daingerfield, *A Vision of the Dawn*
(or *Nude in Grand Canyon*).
Bottom, left to right: Jay Van Everen, *Amerindian
Theme;* H.C. Westermann, *The Stranger;* Roger Brown,
Saguaro's Revenge.
All cover images are details. Full credit information is listed
with full illustration of each image within the catalog.

Library of Congress Cataloging-in-Publication Data

Hough, Katherine Plake.
    Transforming the western image in twentieth century
American art / Katherine Plake Hough, Michael Zakian.
      p. cm.
    "Catalog of an exhibition."
    Includes index.
    1. West (U.S.) in art — Exhibitions. 2. Art, American —
Exhibitions. 3. Modernism (Art) — United States — Exhibitions.
4. Art, Modern — 20th century — United States — Exhibitions.
I. Zakian, Michael, 1957–   . II. Palm Springs Desert Museum.
III. Title.
N6512.H68 1992   91–38799
704.9'49978 — dc20   CIP

ISBN 0-295-97195-9

Catalog Design
Lilli Colton Design, Glendale, California

Typography
Central Typesetting, Los Angeles

Printing
Overseas Color Printing Corporation, Hong Kong

Table of Contents

4   Lenders

5   Foreword
Fritz A. Frauchiger

7   Acknowledgments
Katherine Plake Hough and Michael Zakian

9   Introduction
Katherine Plake Hough and Michael Zakian

11   New Visions of the West, 1905–1951
Michael Zakian

71   A Contemporary Transformation
Katherine Plake Hough

107   Catalog of the Exhibition

110   Index of the Artists

111   Board of Trustees
Palm Springs Desert Museum

Lenders to the Exhibition

An anonymous lady
Arizona West Galleries, Scottsdale
Will Barnet
The Buck Collection, Laguna Hills, California
James and Linda Burrows, Los Angeles
The Butler Institute of American Art, Youngstown, Ohio
Colorado Springs Fine Arts Center, Colorado
Peter Dean
The Denver Art Museum, Colorado
Terry Dintenfass Gallery, New York
Eiteljorg Museum of American Indian and Western Art, Indianapolis
Adolph and Esther Gottlieb Foundation, Inc.
Red Grooms, New York
Katherine H. Haley
The Harmsen Collection
Helen and Marshall Hatch
Hirshhorn Museum and Sculpture Garden, Smithsonian Institution, Washington, D.C.
Ipswich Historical Society, Massachusetts
Dr. and Mrs. William C. Janss, Jr.
Jonson Gallery of the University Art Museum, University of New Mexico, Albuquerque

Phyllis Kind Gallery, New York
Kent M. Klineman
James E. Lewis Museum of Art, Morgan State University, Baltimore
Los Angeles County Museum of Art, California
Mr. and Mrs. John W. Mecom, Jr., Houston
Montana Historical Society, Helena
The Montclair Art Museum, New Jersey
Museum of Fine Arts, Museum of New Mexico, Santa Fe
The Newark Museum, New Jersey
The Oakland Museum, California
Oklahoma City Art Museum, Oklahoma
P.P.O.W., New York
Palm Springs Desert Museum, California
The Gerald Peters Gallery, Santa Fe
Phoenix Art Museum, Arizona
Mr. and Mrs. Alan J. Pomerantz
Portland Art Museum, Oregon
Mrs. Richard Pousette-Dart
Mr. and Mrs. John Pritzker
Private Collections
The Regis Collection, Minneapolis
Michael Rosenfeld Gallery, New York
Roswell Museum and Art Center, New Mexico
San Francisco Museum of Modern Art, California
Santa Barbara Museum of Art, California
Seattle Art Museum, Washington
University Art Museum, University of California, Berkeley
University of Maine at Machias Art Galleries, Machias
University of New Mexico Art Museum, Albuquerque
Walker Art Center, Minneapolis
Frederick R. Weisman Art Foundation
Whitney Museum of American Art, New York

4

Foreword

Over the years many exhibitions have been organized by the curatorial staff of the Palm Springs Desert Museum. There have been notable ones to be sure, but this exhibition *Transforming the Western Image,* posits a thesis opening new territory. Certainly "western" images have enjoyed a plethora of academic investigations. What makes this one unique is the insightful way in which curators Katherine Plake Hough and Michael Zakian have revealed the relationship, decade by decade, between Modernism and the West.

This rich developmental period of American art is well represented with over seventy works of the twentieth century. They have been borrowed from both private collectors and institutions, and I am grateful to these lenders for so graciously lending these pivotal works. Each agreed to a one-year loan covering the time the exhibition will be shown at our Museum and travel to the Boise Art Museum, Boise, Idaho; Tucson Museum of Art, Tucson, Arizona; and The Rockwell Museum, Corning, New York.

The Contemporary Art Council and Cass Graff-Radford, its chairman, deserve praise for providing funds in support of this project. The Board of Trustees and I also offer special acknowledgment to Katherine Plake Hough and Michael Zakian for their consistently dedicated professionalism. This important exhibition and catalog are replete with their resourcefulness and acumen.

Fritz A. Frauchiger
Director

Acknowledgments

An exhibition of this size and scope cannot be accomplished without the assistance of numerous individuals. We are grateful to those who assisted in the organization of this exhibition and publication.

Our deepest gratitude is extended first to those lenders listed elsewhere in this publication who have agreed to share their paintings. They made great sacrifices to loan works of art for the duration of the exhibition tour. We also offer special thanks to the participating artists for their cooperation and for patiently answering our questions.

To many influential museum professionals—museum directors who permitted the inclusion of works from their permanent collections; curators, registrars and other staff members who collaborated with us through complicated arrangements and itineraries—we are especially appreciative. Many individuals provided valuable additional assistance, among them Kathy Reynolds, Registrar, Colorado Springs Fine Arts Center; Dianne Perry Vanderlip, Curator of Modern and Contemporary Art, and Jane Fudge, Administrative Assistant, the Denver Art Museum; Margo Galacar, Curator, John Heard House, Ipswich Historical Society; Tiska Blankenship, Associate Curator, Jonson Gallery of the University Art Museum, University of New Mexico; Alejandro Anreus, Associate Curator, the Montclair Art Museum; Stephanie Turnham, Registrar, Museum of Fine Arts, Museum of New Mexico; Margaret DiSalvi, Librarian, and Audrey Koenig, Registrar, the Newark Museum; Harvey Jones, Senior Curator, the Oakland Museum; Christopher Youngs, Curator, and Jane Hazelton, Registrar, Oklahoma City Art Museum; Nancy Doll, Curator of Twentieth Century Art, Santa Barbara Museum of Art; Vicki Halper, Assistant Curator of Modern Art, Seattle Art Museum; Jane Kamplain, former Registrar, University Art Museum, University of California, Berkeley; Kittu Gates, Registrar, University Art Museum, University of New Mexico; and Ellin Burke, Assistant Registrar, Permanent Collection, Whitney Museum of American Art.

Corporations and foundations have also made specific works available for *Transforming the Western Image.* For their cooperation and participation we are sincerely appreciative for the efforts of Gerald E. Buck, the Buck Collection; Sanford Hirsch and Susan Loesberg, the Adolph and Esther Gottlieb Foundation; William Harmsen and F.C. Hilker, the Harmsen Collection; Vicki Gilmer, the Regis Collection; and Eddie Fumasi, the Frederick R. Weisman Art Foundation.

Galleries and their staffs have been untiringly helpful in tracing specific paintings, contacting private collectors and securing loans. Many have answered readily our repeated requests for photographs and documentation. Special thanks are due Margo Leavin and Lynn Sharpless, Margo Leavin Gallery; Bob Dulak, Terry Dintenfass Gallery; Abe Hays, Arizona West Galleries; Roberta Cove, Rena Bransten Gallery; Dorothy Goldeen, Dorothy Goldeen Gallery; Sandra Starr, James Corcoran Gallery; Ulrike Kantor, Ulrike Kantor Gallery; Ron Jagger, Phyllis Kind Gallery; Wendy Osloff, P.P.O.W.; Dara L.D. Powell and Jackie Cunningham, the Gerald Peters Gallery; Lane Talbot and James Reinish, Hirschl & Adler Galleries; and Michael Rosenfeld, Michael Rosenfeld Gallery.

We are eager to acknowledge many other individuals who have assisted us in a variety of ways, Dee Weldon-White, the Anderson Collection; Ellen Simak, Curator of Collections, Hunter Museum of Art; Marla Price, Chief Curator, Modern Art Museum of Fort Worth; Barbara Trelstad, Registrar, Zimmerli Art Museum; Alison de Lima Greene, Associate Curator of Twentieth-Century Art, Museum of Fine Arts, Houston; Debra Garland, Registrar, Indiana University Art Museum; Julie Campbell, Red Grooms Studio; and Norlene Tips, Curator of Art, Mr. and Mrs. John Mecom, Jr. Collection. Graphic designer Lilli Colton deserves high praise for the handsome design of this catalog, which she successfully supervised through all phases of its production and printing.

We are proud to share the art in this exhibition with audiences in other regions of the United States. The spirit of support extended by the directors and curators of the host museums—Dennis O'Leary, Director, Boise Art Museum; Robert Yassin, Director, Tucson Museum of Art; and Robyn G. Peterson, Curator of Collections, the Rockwell Museum—has been evidenced by their ready willingness to participate in the endeavor.

The Palm Springs Desert Museum staff, whose cooperation and diligence are apparent in both the visible and invisible aspects of the exhibition and its accompanying publication, worked diligently as always to make this undertaking possible. Kathy Clewell, Museum Registrar, assisted with catalog production while dealing with the enormous tasks involved in facilitating loan agreements, insurance certificates and facility reports necessary to travel an exhibition of this size. In addition, she was responsible for the enormity of the arrangements for the nationwide crating and shipping to assure the safe arrival and handling of each painting at each institution. Iona M. Chelette, Art Research Associate, conscientiously attended to numerous details with the catalog organization; Mary Fahr, Museum Librarian, provided the Art Department with an abundance of reference material; William McCracken and James Taylor, Art Preparators, and Douglas Douglass, Exhibit Builder, installed the exhibition with their customary skill and creativity. Also helpful were Barbara Wharton, Administrative Assistant, who patiently typed portions of the catalog manuscript and Suzanne Cunningham, Secretary, who provided the necessary clerical support. Barbara McAlpine, Director of Public Information, edited selected portions of the manuscript, making suggestions with precision and sensitivity. Dr. Janice Lyle, Director of Education, contributed to the success of the project by organizing innovative educational programs and a provocative symposium to enhance the exhibition. Finally we thank Fritz A. Frauchiger, Museum Director, who encouraged us all by his confidence in the exhibition concept and our ability to bring the project to fruition.

The exhibition and publication mark the eighth project funded in part by the Museum's Contemporary Art Council. Grateful recognition is extended to Council members whose support of art programming continues to benefit the Museum and the community.

Katherine Plake Hough
Curator of Art

Michael Zakian
Assistant Curator of Art

Introduction

By the twentieth century the Wild West had ceased to exist. When the frontier was declared closed in the 1890s, there were no new lands to explore, and a chapter in American history had come to an end. Subsequent generations had to come to grips with this legacy. For artists this involved a thorough reevaluation of the role played by the West in our culture.

The West presented a paradox for the twentieth century. It had outlived its popular image and yet continued to haunt people's imaginations. What exactly is the West? What constitutes the western experience? Is it meaningful to speak of such a thing in the modern era?

Some artists remained true to the historic events that shaped the country. These painters continued to create representational images of a way of life that had vanished or was disappearing. Typified by Frederic Remington and Charles Russell at the turn of the century, and by the Taos Society of Artists in later decades, these artists were essentially traditionalists who remained committed to maintaining a picture of the West that in truth was less relevant with each passing year. Their themes continued to be the cowboys, Indians and pioneers long after these figures ceased to play an active role in determining events in the West.

Transforming the Western Image in Twentieth Century American Art addresses another chapter in the history of American art and the West. It focuses on the many ways in which artists have wrestled with the larger meaning and importance of the frontier experience. In the twentieth century certain painters began to explore various aspects of our past, refusing to be limited by narrow stereotypes. They addressed once again the old myths but did so with a new critical spirit. They chose to strip away the veneer of appearance and face the facts of western life.

This exhibition of seventy-one works of art by forty-one artists covers the years between 1905 and 1991. It focuses on the changing relationship of Modern art and the West. It would appear that the very concept of the Old West is antithetical to the principles of Modernism, yet for some artists the myths of the West gave substance to their explorations of innovative painting styles. Many young American artists early in the century were interested in the new art being produced in Europe and wanted to adopt these styles without sacrificing their American roots. The stereotypes of the Wild West allowed some of them to embrace Modern art while retaining a sense of national identity.

There is no single modern view of the West. Decade by decade each generation of artists has found new aspects of the land to investigate and explore. This exhibition traces the ways these artists reevaluated the western experience in terms of contemporary issues.

In the first decades of this century some artists chose to represent the West as a land of peace and plenty much in the manner of classical representations of Arcadia. At the same time, more experimental artists began to borrow elements of European Modernism but chose to use western subjects to give their work an American flavor.

The 1920s through the 1940s saw the full flowering of modernist depictions of the West as painters found a way to render the stark, simple shapes of the western landscape by using the bold, simplified forms of Modern art. The optimism of the turn of the century was shattered in the 1930s by the Great Depression, which had devastating effects throughout the country. Some artists reacted by depicting the blight that spread across the country; others tried to transcend the unpleasant facts of daily reality by turning to pure abstraction.

The 1930s also saw the emergence of a new form of painting in America that was to become known as Abstract Expressionism. This style emphasized inner states of consciousness rather than external reality. For many of these painters examples of Native American art helped them arrive at this revolutionary new style.

After World War II, the myths and visual power of the West continued to inspire a wide range of artistic imaginations. By now the West had become a part of American popular culture. Western themes were important to Pop Art, which drew attention to our nation's myths and stereotypes. From the 1960s through the present, artists have turned to western images for a vast variety of purposes ranging from the humorous to the conceptual. Neo-Expressionists have found in the West subjects of great emotional weight; Postmodern appropriation artists focused on cultural clichés.

No longer seen as an endless horizon, the mystique of the western landscape has survived in art forms that transcend the boundaries of conventional painting and sculpture. Since the 1960s the phenomenon of earthworks resurrected a nineteenth century search for new experiences within the land itself. In a sense, artistic representations of the West have come full circle. From our perspective at the end of the twentieth century, we can survey how artists found different meanings in the western experience.

K.P.H.

M.Z.

9

Arthur B. Davies
The Edge of the Redwood Forest. 1905
oil on canvas
18 x 30 inches
The Regis Collection, Minneapolis

New Visions of the West, 1905–1951

THE IDYLLIC WESTERN LANDSCAPE

ARTHUR B. DAVIES ■ In the early years of the twentieth century there was a strong tendency among advanced east coast artists to look upon the American West as a pastoral realm of milk and honey, a natural utopia. Far from the major cities of the East and untouched by the problems that plagued civilization, it remained pure, the last Eden on earth. This idyllic mood permeates the work of Arthur B. Davies as seen in *The Edge of the Redwood Forest*, which was based on an important trip he made to California in 1905.

Born in Utica, New York, Arthur B. Davies (1862-1928) moved with his family to Chicago in 1878 where he worked at the Chicago Board of Trade for a short period while studying at the Chicago Academy of Design.[1] His first encounter with the West was in 1879 when he visited Colorado seeking relief from respiratory problems, a trip that seemed to have no appreciable effect on his artistic growth. He was more impressed with the art he saw on travels through Europe in the 1890s and developed a poetic, classical style based upon an eclectic array of lyric figural compositions from throughout history. His many sources included Hellenistic art and Pompeiian murals, Giorgione and the poetic landscape tradition of the Venetians, the linear rhythms of Botticelli and the English Pre-Raphaelites, as well as the sensuous classicism of Puvis de Chavannes and Hans von Marées, two late nineteenth-century artists who bridged the gap between academicism and symbolism.

In the early 1900s Davies' original style appeared in the form of gentle, lyric landscapes drawn largely from the imagination and populated with graceful nude figures. These paintings were also influenced by the romantic, glowing color of George Inness's late landscapes and by Whistler's tonalism, a delicate use of atmosphere to create moody, ambient light. These compositions fall into the category of fantasy or caprice and remain some of the finest American manifestations of this genre.

Davies' four-month trip to the West, from June to September 1905, marked a significant shift in emphasis in his art. Judging from the paintings inspired by this trip, he was most interested in the redwood forests and the grand peaks of the Sierra Nevada. The West did not alter his subject matter, which was consistent, even repetitious, for all of his life. It did provide him with a new sense of grandeur and an opportunity to place his imaginative figure groups in a real setting.

To Davies, the California landscape represented a new Arcadian realm, possessing a natural grandeur that rivaled the noblest monuments of the ancient world. *The Edge of the Redwood Forest* is friezelike, duplicating the shallow space and lateral disposition of forms seen in Greek and Roman architectural art. He chose an unusually low vantage point — from ground level — in order to make the composition seem architectonic. The trees stand as straight as marble columns; in fact Davies utilized them as surrogate columns with bark substituting for the fluting of the classical orders. Although nothing man-made appears in the composition, Davies nevertheless managed to create an allusion to classicism by drawing a parallel between the majesty of the redwoods and the splendor of ancient architecture.

Davies' earlier works were highly artificial, with nymphs dancing aimlessly across a stage with no sense of purpose. In *The Edge of the Redwood Forest*, however, a narrative element appears that gives a clearer moral purpose to the image. The two female nudes leading a horse into the grove of trees point forward, offering direction for the others. They gesture, trying to coax their cavorting friends to follow. They point down a road that leads deep into the forest, a path with

11

strong symbolic allusions to the future. Deep within the forest one can barely discern the minute forms of people standing amidst towering redwoods, bathed by the broken light filtering through dense foliage. They are enveloped in a quasi-religious light. Instead of adhering to the classical ideal of clarity, Davies gives us a modern, spiritual image of people dissolving into harmony with nature's greatness.

Frank Jewett Mather, Jr., an important newspaper critic in New York in the early 1900s, wrote that the 1905 trip to the West helped broaden Davies' art, that the artist began to include in his paintings "those vast mountain panoramas inspired by the Sierras. The little Arcadia of the early pastorals had become gigantic, cosmic."[2] Davies' experience in California helped instill a new sense of dignity in his work.

ELLIOTT DAINGERFIELD ■ Arthur Davies was not the only artist to employ allegorical nudes in his depictions of the American West. Elliott Daingerfield's *A Vision of the Dawn* of 1913 is similar to Davies' work in subject matter. It depicts a female nude standing at the rim of the Grand Canyon in Arizona. But the mood and, more important, the ideal represented by the work of the two differed dramatically.

Daingerfield (1859-1932) was a close contemporary of Davies, but his art belongs in sensibility to a preceding generation — that of America's Gilded Age. The closing decades of the nineteenth century brought great prosperity to the United States. Rapid industrialization in the years after the Civil War produced large fortunes and an increasing awareness of a need to create a cultural tradition equal to our newfound economic importance. To meet these demands there appeared the "professional artist," an individual of natural skill and fine training who assumed the task of channeling various cultural ideals into appropriate visual images. These artists, of whom Daingerfield was a prime example, saw themselves as upholders of our classical heritage and of "official" culture, combining learning with taste to produce dignified images that were edifying, moralizing and uplifting. Contemporaries of the French Beaux Arts academicians, they proudly declared a role as singular champions of high culture. More often than not, these arbiters of taste found their prized values — truth and beauty — in empty recreations of High Renaissance and Baroque allegories.

A native of North Carolina, Daingerfield spent most of his career painting there and in New York, where his work received acclaim in the first decades of the twentieth century. Based on various styles sanctioned by current taste, his style was eclectic. Instead of synthesizing various manners into a style of his own, he tended to pay homage to a particular master in individual works, producing one canvas in the

manner of Titian, another after Raphael. His taste was not limited to the Renaissance. He also produced more personal works based on such American masters of the late nineteenth century as Winslow Homer, Albert Pinkham Ryder and Ralph Albert Blakelock.

Daingerfield first saw the Grand Canyon in 1911. After being commissioned by the Santa Fe Railway to paint the natural wonder to entice travelers west, he executed a number of canvases between 1911 and 1915, all true to his reputation for idealizing his subjects. He produced a number of straightforward landscapes of the canyon distinguished by their glowing, opalescent color. Of greater interest are his allegorical canvases, which depict idealized nude figures at the edge of the canyon. In *A Vision of the Dawn* a female nude stands next to a mist-enshrouded Grand Canyon. The title makes a clear reference to the dawn of time, with the human figure emerging from the morass of nature below. In a similar painting, *The Genius of the Canyon*, also of 1913, Daingerfield depicts a reclining female nude overlooking the canyon, with a mystical white city rising in the distance. He wrote a poem to explain the image:

> Strip from the earth her crust
> And see revealed the carven glory of the inner world.
> Templed — domed — silent: —
> The white Genius of the Canyon broods.
> Nor counts the Ages of Mankind
> A thought amid the everlasting calm.[3]

He saw in the canyon an equivalent to the carved white marble wonders of the ancient world. The outcroppings of rock were to him like the domed temples of the past. He saw the Grand Canyon metaphorically as something that lasts throughout the ages, a deep fissure allowing a look deep into the earth's history. The allegorical nude is defined as a "genius," which is a spirit or personification. She is a brooding, contemplative figure who calmly sits beyond the touch of time.

In another painting, *The Sleepers* of 1914, Daingerfield depicts a group of nudes languorously nestled among the rocks at the canyon's edge, awakening from a long sleep. In the poem accompanying this work, the artist wrote:

> Age on age the Sleepers rise
> To see in dreams the canyons splendor rise
> Height from river bed to golden crest
> Gods are they — as you and I, —
> Who see in spirit what eyes deny.[4]

The poses of these figures were inspired by Michelangelo's figures of *Night* and *Day*, and *Dawn* and *Dusk* from the Tombs of Lorenzo and

Elliott Daingerfield
A Vision of the Dawn (or Nude in Grand Canyon). 1913
oil on canvas
48 x 36½ inches
Collection of James E. Lewis Museum of
Art, Morgan State University

13

Giuliano de' Medici. From the references in these poems it is clear that Daingerfield thought of his figures as embodiments of the loftiest ideals. They represent qualities of truth, history, religion and purity.

In these landscapes Daingerfield holds true to the academic tenet that the mind is superior to the eye. He was inspired by the Grand Canyon, but its natural beauty was not enough. He felt he had to improve upon it and give it deeper meaning, in order to produce a significant work of art. He favored using the imagination to improve upon nature and once said, "if one must paint from nature, paint from it meaning away from it."[5] Daingerfield used the term "synthetic" to refer to landscapes such as these, inspired by nature but created within the imagination.[6] The ideal of synthesis guided the development of more avant-garde currents in Europe at this time, leading to the 1912 development of Synthetic Cubism in France. Daingerfield's idea of synthesis was far less radical; he believed it was best represented in the work of George Inness in which various elements drawn from nature are recombined into a harmonious whole. The glowing color in *A Vision of the Dawn*, painted with many layers of glazes, indicates a clear debt to the luminescent, poetic landscapes of Inness's late years.

It is worth noting the degree to which Elliott Daingerfield's work diverges from Davies'. Both used the nude allegorically to project notions of purity, beauty and intelligence. But Daingerfield's figure is only partially nude. The cloak she carries over her shoulder signifies that she represents two worlds: culture with its strict moral codes and nature with its freedom from inhibition. There is an awkwardness to the figure as she self-consciously covers herself by crossing her arms over her chest. Her discomfort represents the self-consciousness of the time, when people voiced admiration for the ideals of the classical past but still looked upon nudity with shock.

In Davies' intellectual paintings there is a frivolousness bordering on decadence. His listless figures seem too delicate and sensitive to face the rigors of the real world. Daingerfield, on the other hand, opposes the indifference and fragility of Davies' nymphs with a robust figure who stands firmly on the ground, self-possessed and self-assured. She is an embodiment of lofty human reason, someone untouched by the base aspects of nature. She signifies Daingerfield's belief that mankind's noblest state is one of reason, never to be tainted by elements of the natural world.

Daingerfield was typical of the academic artists of the late nineteenth and early twentieth centuries who believed that art should provide moral direction for society. He found in the Grand Canyon a natural grandeur that he felt should serve as a model for civilization. He turned the reality of landscape into fiction to indicate how people have lost touch with a nobler past. His vision was dated and was soon superseded by more advanced styles.

ARTHUR WESLEY DOW ■ While artists like Davies and Daingerfield adhered to a nineteenth-century use of allegorical nudes, others at the time employed more modern forms of composition to capture the awe-inspiring spectacle of the western landscape. Arthur Wesley Dow painted *Bright Angel Canyon* in 1912, just one year after Daingerfield's *A Vision of the Dawn*, but it represents a vision radically new in its freshness and directness.

Arthur Wesley Dow (1857-1922) was a decorative landscape painter who had an enormous impact as an educator at the turn of the century. He championed an appreciation of Oriental art and design at a time when that aesthetic was still novel in America. Through a successful book, *Composition* of 1899, and through his important position as the head of the art education department of Teacher's College, Columbia University — which he assumed in 1903 — he had a profound influence upon the art and aesthetics of the American Arts and Crafts period. His most famous student was Georgia O'Keeffe, who used many of his principles to create her own style.

Originally working in an uninspired Barbizon style, Dow was known to have been in contact with Gauguin's symbolist circle at Pont-Aven in Brittany in 1889, however their primitivist art seems to have had little influence upon him. A turning point occurred in 1891 when he met Ernest Fenollosa, curator of Oriental art at the Museum of Fine Arts in Boston. Fenollosa introduced Dow to Oriental methods of composition, including shapes conceived as simple masses, line used as an evocative end in itself and *notan*, a Japanese method of composing in value contrasts of light and dark.

In most of his mature work, Dow's subjects consisted of simple landscapes featuring picturesque genre elements from his native Ipswich, Massachusetts, a low lying marshland. Most of his compositions were balanced asymmetrically, in the manner of Japanese woodblocks, and featured a slow-moving sinuous line. They tended to adhere closely to his principles of composition set forth in his teachings. In the first few years of the twentieth century, Dow experienced a certain dissatisfaction with the redundancy within his art and looked to find a solution in the more open, energetic brushwork of Maurice Prendergast. At the same time he wanted to paint something grand enough to add substance and weight to his essentially decorative style. He was tiring of his lowland landscapes, wanting to paint "some of the *big things* of the world."[7] He found this subject in the Grand Canyon.

Dow visited the Grand Canyon in Arizona in 1911 and 1912 making meticulous site sketches that he augmented with photographs and written descriptions. *Bright Angel Canyon* of 1912 may be the finest of his Grand Canyon scenes. A swirling movement of broken brushstrokes in the middistance animates the entire scene. In a written statement that accompanied an exhibition of the Arizona paintings at

Arthur Dow
Bright Angel Canyon. 1912
oil on canvas
30 x 40¼ inches
Collection of Ipswich Historical Society
Photography by Greg Dann

the Montross Gallery in April 1913, Dow wrote that this painting was inspired by a snowstorm he experienced in late November. The complex light effects resulted from the turbulence of the air within the canyon and from the fact that the snow melted as it fell to the lower depths, changing the refraction index and the resulting quality of light.

In his introduction to the exhibition catalog, Dow explained his fascination for the Grand Canyon:

> You ask what attracted me to the Grand Canyon so far from my New England marshes. Color first of all — color "burning bright" or smoldering under ash-grays. Then, line — for the color lies in rhythmic ranges, pile on pile, a geologic Babylon. This high, thin air is iridescent from cosmic dust: shapes and shadows seen in these vast distances and fearful deeps, are now blue, now liberating with spectral hues. At sunset the "temples" are flaming, red orange — glorified like the Egyptian god in his sanctuary.
>
> The Canyon's color and line cannot be well-expressed without study of the structure, for this is neither "chaos" nor "hell" but orderly world-building. . . .
>
> The Canyon is not like any other subject in color, lighting or scale of distances. It forces the artist to seek new ways of painting — its own ways. Its record of the world's beginning holds for us the romance of geology.[8]

The persistent romantic characterizations indicate Dow's indebtedness to a nineteenth-century attitude that equated meaning in art with the highest and noblest themes, as we saw in Elliott Daingerfield's work. When looking upon the Grand Canyon, he saw not just nature but a metaphor for mankind's universal quest for order and purpose. To Dow, the canyon represented "orderly world-building" and a "geologic Babylon," an earthen wonder equal in scope to the grandest of ancient cultures. The buttes and outcroppings of rock were "temples," the air filled with "cosmic dust."

Dow's art was far from being modernist for it attempted to overthrow no conventions or establish a new sense of purpose for art. It was modern, or forward-looking, in championing the artist's individual intuition as the basis for good composition. This notion of Modern Art was popular at the turn of the century. Based on the idea of stylization, the approach had its base in the decorative Art Nouveau style and sought to improve upon nature by making it more balanced and harmonious. It was palatable to the public at large because it did not involve a drastic break with traditional values. Rather it retained all the elements of conventional taste and improved upon them by redesigning and simplifying pictorial relationships.

Dow's *Bright Angel Canyon* of 1912 is conventional in the disposition of forms in space. Although its basic structure is based on that of traditional representational painting, the vigor of the image reflects the awe he felt upon viewing the spectacle. There is a strong spiritual rather than pictorial quality to his brushwork. Instead of capturing a specific effect of natural light, it swirls throughout the composition, conveying a sense of resonant energy. The brushstrokes seem to radiate from a spot of light at the center of the horizon. This other-worldly force seems to sweep forth to envelop all before it. The symbolism is straightforward: the Canyon is a source of life energy, a place of renewal. The titles Dow gave to the seventeen paintings that constitute the Grand Canyon series are usually descriptive, but also include references to spiritual and utopian themes. *Unpeopled Cities* and *Cosmic Cities* convey Dow's vision of the canyon embodying the promise of a new and pure social order. *The Glory of Shiva* is a direct reference to the Eastern spirituality that he believed permeated the place.

It is ironic that the exhibition of these paintings at the Montross Gallery in New York should occur just one month after the close of the Armory Show, the first large-scale exhibition of Modern Art seen in America. Although Dow championed a modern approach in art, he remained opposed to the free experimentation advocated by the most radical of the first-generation European modernists and those Americans intrigued by the new style.

PAUL BURLIN ■ Paul Burlin (1886-1969) is important to the history of Modern Art in the West because he was the first of those to exhibit in the Armory Show to actually visit and paint in the American Southwest. With no formal training in art, he began working at the age of seventeen as a layout artist for the *Delineator*, a magazine of fashion and culture edited by Theodore Dreiser. After taking up painting in his early twenties, he was asked to exhibit in the Armory Show at twenty-seven. At the close of the exhibition in 1913 he visited the Southwest where he remained for several years. Unfortunately the facts of his stay in the West have been lost with time. *Grand Canyon*, painted sometime between his first visit to the Southwest in 1913 and when he finally left the region in 1921, moves away from the realist and impressionist styles popular at the time.

Burlin was keenly interested in and responsive to the latest styles in art, possibly because his involvement with the *Delineator* made him aware of fashionable trends. His art in the years after the Armory Show could best be described as a lyric expressionism that drew upon different currents in Modernism. This merging of disparate styles is seen in *Grand Canyon*: his composition as traditional as Dow's *Bright Angel Canyon*, except that Burlin augmented his rather conventional drawing with unexpected, arbitrary coloring and heavy, dark outlines. He was combining the studied pictorial structure of Cézanne with the freely improvised color of the Fauves.

In this work, unnatural color and a simplification of forms is used to "modernize" what is essentially a straightforward realist composi-

Paul Burlin
Grand Canyon. n.d.
oil on canvas
20 x 25 inches
The Harmsen Collection

tion. Burlin kept the same spatial relations and the same basic grada-
tions of value present in the actual scene. His purples were meant to
show free use of color, but these tones are dull and chalky, never as
vibrant as in the Matisse and Fauve paintings he admired. His art
underscores problems encountered as ambitious young American
painters tried to emulate the new European art. Modernism came to
America secondhand. When it first appeared in the West, far from the
original examples of modern art that were still novelties in the East, it
appeared even more derivative. Burlin, like many American artists of
the time, missed the subtler aspects of this art and failed to use modernist techniques for any consistent purpose.

One thing new in Burlin's work is the way he uses the large scale
of the western landscape to help simplify his imagery. In *Grand
Canyon* he exploited the great width and depth of the canyon to
organize his masses into planes that clearly express spatial location
without resorting to a detailed delineation of each rock and crevice.
Later in the 1920s and 1930s, American artists would further utilize
the breadth of the western land to help them abstract from the forms
of nature.

Maynard Dixon
The Ancient. 1915
oil on canvas
19¼ x 11½ inches
Collection of Katherine H. Haley
Photography by William B. Dewey

18

THE NATIVE PEOPLES OF THE WEST

MAYNARD DIXON ■ A native Californian, Maynard Dixon (1875-1946) began his career as an illustrator and moved to New York in 1907 to be near the publishing industry. He was not particularly fond of life in the East and longed to return to the West. To his friend Charles Lummis he wrote, "I am being paid to lie about the west. I am going back home where I can do honest work."9 He returned to San Francisco in 1912. More a bohemian than a modernist, Dixon always promoted unconventional views but did so within the boundaries of traditional realism.

In 1915 Dixon made a trip through Arizona, stopping at San Carlos, Fort Apache and the Grand Canyon. He produced a number of important paintings of Native Americans, including *The Ancient* of 1915 and a larger variant of this composition, *What an Indian Knows* of the same year. In both these works, an adult Native American male stands alone and motionless in an empty landscape. *The Ancient* is not truly modernist in style yet it embodies a modern outlook in a rejection of anecdote. Instead of placing the figure as part of a more conventional narrative, Dixon uses him as a living icon, a solitary symbol. This type of image has precedents in late nineteenth-century Symbolism. As in Symbolist art, Dixon uses a single static form charged with meaning. Within this figure the artist perceived not just an individual but an entire people and a way of life.

Dixon once said, "The Indian is enigmatic and so he fascinates us, he does away with the superficialities and comes straight to the point. But the most peculiar and subtle thing about him is his faculty of using silence as a means of communication."10 For Dixon the Native American deserved praise for stoicism. He displayed self-control and solemn dignity. Dixon's comments on another of his canvases applies equally well to *The Ancient*: "He is a sage, calm Indian who stands against his own background of mountains, from which he draws his health, wealth, religion and pattern of living. While we get panic-stricken over 'the Market' the Indian puffs his pipe and looks at the sky. The West is big, slow, grand country, with a style all its own. We scurry around it like nervous ants and scarcely take time to let it seep into our souls."11

The bright colors of *The Ancient*, particularly the warm yellows contrasted with the cool pink-purple shadows, derives from Post-Impressionism, which was then still a fairly radical style. Although his drawing is traditional in its delineation of solid masses, the color shows his desire to intensify the image. Although not truly modernist, this painting is advanced in the radical simplification of composition, the elimination of detail and the reduction of forms to simple masses. Dixon was paradoxical in his relation to Modernism but in many ways his stance embodies deep currents that ran through American art of midcentury. He freely borrowed particular formal elements

from Modern Art but refused to embrace the ideology of Modernism. He was later to become aware of modernist currents in New York but rejected what he called the "hot house atmosphere and fake modernisms" of the east coast art establishment: "After listening to exploiter Stieglitz expatiate, and observing so much cleverness and futility, I was glad to quit that stale-air existence and come West."12 The West provided the opportunity to work apart from the intellectual demands of the established art world. It offered Dixon an opportunity to embrace subjects dear to him, such as personal freedom. He once said, "I do not paint Indians . . . merely because they are picturesque objects, but because through them I can express that Phantasy of freedom and space and thought, which will give the world a sentiment about these people which is inspiring and uplifting."13

MARSDEN HARTLEY ■ While Dixon saw the Indian as a solitary, stoic figure, possessing great inner wisdom and strength, Marsden Hartley (1877-1943) chose to see the Indians in another light. He was interested foremost in their culture and especially in their use of ritual as a means of social organization. Hartley was born in Lewiston, Maine, and studied art at the Cleveland School of Art in 1892 before arriving in New York where he enrolled in the National Academy of Design in 1900. After returning to Maine, he began his involvement with Modernism around 1908 with a series of mountain landscapes inspired by the dark, brooding romanticism of Albert Pinkham Ryder and the Divisionism of Giovanni Segantini. In 1912 he traveled to Europe and spent two years between 1913 and 1915 mainly in Germany where he was influenced by the spiritual abstractions of Vassily Kandinsky and Der Blau Reiter, a mystical German Expressionist movement. He began to produce improvisational abstractions in the manner of Kandinsky.

Although Hartley readily adopted the abstract vocabulary of Kandinsky's art, he did not seem to be able to find a theme or purpose in his work. As a means to solve this problem he turned to a series of semiabstractions based on American Indians such as *Indian Composition* of 1914. Having never visited the West, he used this subject as a way to assert his identity as an American, referring to this series as his *Amerika* paintings. He admired the Native American for living in harmony with the land:

> He is the one man who has shown us the significance of the
> poetic aspects of our original land. Without him we should still
> be unrepresented in the cultural development of the world. . . .
> He has indicated for all time the symbolic splendor of our
> plains, canyons, mountains, lakes, mesas and ravines. . . . He
> has learned throughout the centuries the nature of our soil and
> has symbolized for his own religious and esthetic satisfaction all
> the various forms that have become benefactors to him.14

Hartley based his compositions on examples of Native American art he had seen in Berlin's ethnographic Museum für Völkerkunde, and derived his imagery from a variety of Native American works of art from both Pueblo and Plains peoples. He was drawn in particular to the symmetry seen in Indian art. Gail Scott wrote that, "the Amerika series . . . utilize one of the dominant features of Indian design, seen in everything from beadwork to weavings — symmetry. Ideally suited to Hartley's preference for formal, triangular-based formats, symmetry became an important part of his compositional vocabulary."[15] Hartley found in symmetry a comforting sense of order and stability.

The *Amerika* paintings differ from their sources in one fundamental aspect. They are far denser in their organization than any example of tribal art, emphasizing a tight packing of forms into a shallow space. The compositions are crowded with shapes uncomfortably overlapping one another. In Native American art, each symbol is understood to maintain a direct, almost magical, connection to the object it symbolizes. The integrity and individuality of each image is respected as if

Marsden Hartley
Indian Composition. 1914
oil on canvas
47¾ x 47¾ inches
Collection of Vassar College Art
Gallery, gift of Paul Rosenfeld
Photography by Douglas Baz

20

the actual thing itself. Although Hartley greatly admired the Native Americans' oneness with the land, he could only copy their symbols and never duplicate their system of values that was so foreign to Europeans. He lacked the simplicity and clarity of vision of the Native American artist, and used overburdened forms to make up for a lack of incisiveness. He felt it necessary to employ a multitude of abstract symbols, where the original artist needed only a few to convey his message. The space the Indian artist left around his forms revealed his fundamental respect for the thing depicted, and his certainty of its place within the world at large.

ARTHUR B. DAVIES ■ The people in Davies' paintings, even in work immediately following his 1905 trip to California, were always Caucasian. Although profoundly moved by the majestic country in the West he was indifferent to its native inhabitants. He did introduce Native American subjects in *Indian Fantasy,* circa 1918. Davies was the president of the organizers of the Armory Show. An eclectic artist with a deep interest in intellectual currents, he remained open to modern art and gradually altered his style in the middle of the 1910s in the direction of Cubism. He employed this style in a conservative way. Instead of using geometric elements to create new pictorial space, he used them decoratively to enliven traditional drawing.

In *Indian Fantasy*, the figures are contained by a conventional outline much as in Davies' earlier work. What is new is the almost prismatic fracturing of color. Color is applied arbitrarily in small sections independent of each form. In some of his Cubist compositions, the color is bright and glowing. In this canvas, it is somber and keyed to low-value earth tones, offset by small areas of brighter orange and blue.

The subject is seven male figures, nude save for loincloths, who strike various poses while drawing their bows. Behind them one can discern the monochromatic shape of a running horse seen in profile, repeated in black, white, red, yellow and blue. These hunters are not engaged in any real act; Davies used these twisting, bending figures to meditate upon the poetics of movement. Their actions seem more orchestrated than real, more dancelike than practical. Frank Jewett Mather, Jr. found that in general Davies' multifigure compositions grow "out of some sharply bent figure, from which others are derived in natural relation of repetition, reciprocating or opposing thrust."[16]

The figures in this painting are men, unusual in that Davies almost always used women in his work. Perhaps he felt that white European men were too distant from nature to symbolize a harmony with its rhythms, but that Native Americans were free from the inhibitions that limited Caucasian spontaneity. The accumulated energy of this central massing of figures is denser than in Davies' other paintings, and more masculine in keeping with the vital strength of the subject.

Arthur B. Davies
Indian Fantasy. ca. 1918
oil on canvas
17⅝ x 16⅛ inches
Collection of The Newark Museum,
bequest of Miss Lillie Bliss, 1931
Photography by Armen

WALT KUHN ■ Walt Kuhn's *Commissioners* of 1918 is so similar to his friend Arthur Davies' *Indian Fantasy* that it is probable the two were looking at each other's work. Walt Kuhn (1880-1949) moved to San Francisco in 1899 from his native Brooklyn to take a job as a newspaper cartoonist. He took this opportunity to travel through the West, returning to New York in 1900 where he found work drawing cartoons for a number of popular magazines. He served as executive secretary of the Armory Show and traveled to Europe in search of appropriate examples of the new art, reporting his findings to Arthur Davies.

In 1918 he began a series of paintings entitled *An Imaginary History of the American West*, which he worked on until 1923. These highly fictionalized works depict the West in terms of popular stereotypes. For the most part the figures are melodramatic caricatures of standard western characters.

Commissioners is one of the earliest works in the series and also one of the less comic. The work depicts a meeting of U.S. Army officers and Indian chiefs, two groups of commissioners representing their people. The background is completely abstract, consisting of random spots of oranges and browns that deny any illusion of space. The subject is one of confrontation if not outright conflict; the two groups proudly face one another with no sense of sentimentality or exaggeration. The majority of the paintings from this series is less serious and more slapstick in portrayals of humorous characters and rollicking, bawdy situations. For the most part, these paintings are pure fictions bordering on farce.

John Quinn, a lawyer who assembled an important collection of modern art in the years after the Armory Show, partially through Kuhn's guidance, wrote a letter to a New York newspaper praising *An Imaginary History of the American West*:

> They give us back a vision of a vanished life that was not all comedy or all tragedy, but something strangely mixed. . . . There they are, the fellows of Boone and Crockett and Houston and of Deadwood Dick and Calamity Jane and Wild Bill, and the young Buffalo Bill and the young Custer and the soldiers of his day, and all of the adventurous and brave men of those days in the West. They give us a glimpse of the vanished life of the West and of that life's vanished romance. [17]

Quinn seems to have accepted at face value the romantic, nostalgic dimension of Kuhn's fanciful images. He never raised the question of accuracy, preferring popular misconceptions to historical truth.

Walt Kuhn
Commissioners. 1918
oil on canvas
10⅛ x 16 inches
Collection of
Colorado Springs Fine Arts Center,
gift of Vera and Brenda Kuhn

22

THE GLOW OF LIGHT

RAYMOND JONSON ■ For most of his life, Raymond Jonson (1891-1982) was a tireless champion of abstract art in New Mexico, where he lived from 1924 until his death. Drawn to the spiritual aspects of art, he became a founding member in 1938 of the Transcendental Painting Group, a short-lived association of Southwest painters involved with nonobjectivity and the absolute.

His interest in both the transcendental and the West manifested itself while he was still living in the Midwest. Born in Iowa, Raymond Jonson lived in many places from the Midwest through the Northwest before the family settled in Chicago in 1910 where he began study at the Chicago Academy of Fine Arts and the Art Institute of Chicago. He developed an interest in Modern Art after seeing the Armory Show which was exhibited in Chicago. His painting *Light* of 1917 stands out within his oeuvre because it shows, at an early date, his interest in the dramatic spectacle of the sun over a Southwest landscape. The almost theatrical presentation of the image points to the artist's experience in the theater.

In the years between 1913 and 1917 Jonson worked for the Chicago Little Theater as a designer in costume and set design as well as lighting. By creating sets based on clear planar relationships, he learned to think of pictorial design in terms of abstract arrangement. This respect for the rational disposition of forms in space remained with Jonson throughout his life, influencing an interest in finding abstract relationships within natural landscape phenomena.

Jonson made numerous trips from Chicago to the West in the 1910s, visiting Texas, New Mexico and Colorado in 1914, Portland and Colorado in 1917 and Wyoming, Utah and Idaho in 1918. *Light* was inspired by these journeys. This popular work was exhibited throughout the Midwest in the early 1920s and received favorable notices. One newspaper reporter said that, "a Jonson rock has its roots in the dawn of time. His sky is the glorious thing the first beholding eye of man grew accustomed to. His sun is a thing of infinite radiance."[18] The language of this description is similar to that used by Daingerfield and Dow; the West is seen as a timeless realm of majestic proportions. Actually Jonson was interested more in the transcendental qualities of light than in its function as a timeless symbol. He saw light as an abstract force and not simply a natural phenomenon or a hackneyed symbol.

GEORGIA O'KEEFFE ■ Georgia O'Keeffe (1887-1986) was another artist profoundly moved in the 1910s by the experience of the light on the western landscape. She is generally regarded as one of this country's greatest painters, a reputation based to a large extent upon her paintings of the American desert, a stark, arid region she renders as sensuous. These paintings of the New Mexico landscape, which she produced from the 1930s until her death, stand as icons of modern American painting.

O'Keeffe made her first trip to the West in late August 1912 when she assumed a post as supervisor of drawing, overseeing art classes in half a dozen schools in Amarillo, Texas. She stayed until the spring of 1913. In September 1916 she returned to the state, taking a job as art supervisor for West Texas State Normal School, a two-year teacher's college in Canyon, Texas, twenty miles south of Amarillo.

Unlike Daingerfield and Dow, who saw the West with a rhetoric derived from classical literature, O'Keeffe drew upon perceptions that were strongly individual, infused with personal emotions. She recalled that as a young girl "Texas had always been a sort of far-away dream" and that she "had listened for many hours to boy's stories . . . stories of the Wild West, of Texas, Kit Carson and Billy the Kid. It always seemed to me that the West must be wonderful—there was no place I knew of that I would rather go—so when I had a chance to teach there—off I went to Texas—not knowing much about teaching."[19]

When she arrived in Texas, her first impression was that she "couldn't believe Texas was real. When I arrived out there, there wasn't a blade of green grass or a leaf to be seen, but I was absolutely crazy about it. . . . For me Texas is the same big wonderful thing that oceans and the highest mountains are."[20] Clearly she was taken not by allusions to the ancient world but by its bigness, its openness, and in a more abstract sense, its romance. The plains of the northern Texas panhandle are much flatter than the prairie O'Keeffe knew in her native Wisconsin, offering a viewer the spectacular contrast of the horizontal land and the blue sky above. In writing to her friend Anita Pollitzer, she said that "there was nothing but sky and flat prairie land—land that seems more like the ocean than anything else I know. . . . I am loving the plains more than ever it seems—and the SKY—Anita you have never seen SKY."[21]

The elements also seemed more real as there was nothing to hinder the wind blowing across the ground. Sunrises and sunsets were particularly dramatic, and she used to rise before dawn to watch the sun climb up from the dark eastern horizon. Sensitive to subtle light effects, she had noticed that "the light would begin to appear and then it would disappear and there would be a kind of halo effect, and then it would appear again."[22]

When she moved to Canyon, O'Keeffe had been working in black and white, favoring charcoal as a way to work out new designs without being troubled by questions of color. In Texas she began to introduce color again. Her first choice was blue, which she used in monochromatic watercolors influenced by the sky. Using one color at a time, she explored yellow, green and finally red before she returned to a full spectrum of hues in her work.

24

Raymond Jonson
Light. 1917
oil on canvas
45 x 42 inches
Collection of Museum of Fine Arts,
Museum of New Mexico, gift of
John Curtis Underwood, 1925

Georgia O'Keeffe
Red and Black . 1916
watercolor
12 x 9 inches
Courtesy of The Gerald Peters Gallery,
Santa Fe, New Mexico

Red and Black of 1916 is one of these predominantly red paintings. O'Keeffe's Canyon, Texas, watercolors emphasize a repetition of organic, curving forms. The pattern of forms conveys the unrelenting constancy of the environment. Although modest in size, this painting captures the immensity of the plains. O'Keeffe wrote to a friend with excitement, saying that she didn't know best what she liked better, the mountains or the plains. "I guess it's the feeling of bigness in both that carries me away."[23] She also said that "the plains — the wonderful great big sky — makes me want to breathe so deep that I'll break — There is so much of it — I want to get outside of it all — I would if I could — even if it killed me."[24]

She could only capture this sense of overwhelming scale using abstract forms. She made the bold step to break with description and paint sensation, not just appearance. This was an important step in her development as she finally found a subject that allowed her to reject the influence of other artists and begin looking within.

Interestingly, she received her first push in this direction from Arthur Dow who was her instructor at Teacher's College, Columbia University from 1940 to 1916. O'Keeffe said, "I think it was Arthur Dow who effected my start, who helped me to find something of my own. . . . This man had one dominating idea: to fill a space in a beautiful way — and that interested me."[25] To fill a space in a beautiful way meant that you had to fill the *whole* space. O'Keeffe began look-

ing for the large form, the large sensation that would allow her to do this. Before visiting Canyon, Texas, she did a number of transitional works where she enlarged simple objects — such as a rooster or the side of a small house — to fill the whole page, but the subjects themselves didn't warrant this attention, and the results seemed forced. In Texas she discovered the largeness that she sought to convey in her paintings.

Dow's teaching methods were revolutionary for they broke with the then current method of painstakingly copying nature or the work of old masters. Rejecting this slavish attention to detail, Dow offered exercises in design that even students unskilled in drawing could practice. Many exercises had to do with dividing squares into subdivisions of interesting proportions or balancing a simple mass with smaller abstract shapes. O'Keeffe was aware of Dow's paintings of the Grand Canyon and said, "I can understand Pa Dow painting his pretty colored canyons — it must have been a great temptation — no wonder he fell."[26]

O'Keeffe fell under the spell of the West as well but with different results. She began to move in the direction of abstraction. Focusing on the feeling and sensation of the Texas landscape, she gained the means to paint subjective impressions of the land. For her, objective reality was infused with the subject's individual response to it. This attitude would become important in modern paintings about the West.

STUART DAVIS ■ A number of artists during the first two decades of the twentieth century showed an interest in Modernism, but, with the exception of the isolated experiments of Marsden Hartley, Georgia O'Keeffe and a few others, their efforts fell short of attaining an original modern vision. What was needed was a radical shift in point of view and an entirely new set of values in art. One person who represented this new outlook was Stuart Davis. His involvement with the West was short-lived, but a new perspective emerges in the few works he did paint.

Stuart Davis (1894-1964) was a transitional figure. He bridged the gap between the old Ashcan School of gritty urban realism and the new world of abstract color and form. The son of a Philadelphia newspaper art editor, Davis studied with Robert Henri, charismatic leader of the Ashcan School, from 1910 to 1913 and worked with John Sloan as a cartoonist and illustrator for the left wing journal *The Masses* from 1913 to 1916. Unlike other artists of Henri's circle, who remained loyal to the idea of socially committed realist art, Davis was profoundly moved by the Armory Show. He readily adopted the principles of abstraction, maintaining that art was a formal arrangement of pure pictorial motifs.

At the urging of John Sloan, Stuart Davis visited New Mexico during the summer of 1923. Although his stay lasted only about a season and his production was limited to a small number of paintings, including *New Mexico Gate* of 1923, this period marked a turning point in the history of modern American art and the West. Davis did not like New Mexico. His recollections of the time spent in Taos dwell on the shortcomings of the place:

> Just as John Sloan used to rave about Gloucester, and I went there on his recommendation, it was the same about New Mexico — but with a difference. I spent three or four months there in 1923 — until late fall — but did not do much work because the place itself is so interesting. I don't think you could do much work there except in a literal way, because the place is always there in such a dominating way. You always have to look at it. Then there's the great dead population. You don't see them but you stumble over them. A piece of pottery here and there and everywhere. It's a place for an ethnologist not an artist. Not sufficient intellectual stimulus. Forms made to order, to imitate. Colors — but I never went there again.[27]

What Davis did not like about the land is significant for it draws attention to the profound shifts in sensibility that took place in the 1920s and after. When he said he did not do much painting in New Mexico because it was "so interesting," he meant that it was *too* interesting. It demanded too much of his attention, not leaving enough time to concentrate on the formal problems of art. At the same time, Davis despaired that Taos did not offer "sufficient intellectual stimulus." Raised in an urban environment, he had an avid interest in modern technology. He firmly believed that if his art was to have relevance for his time, it had to reflect the profound changes brought about by the radio, automobile and airplane. These inventions altered the pace of modern life, but in New Mexico their impact was negligible. Horse-drawn wooden carts were as popular there as they had been for centuries. The quaint, quieter mode of life held little appeal to someone like Davis who liked to frequent New York City's jazz clubs and feel the pulse of the urban environment.

Fortunately, we have records of Davis's opinions about art at the time. In March 1923, just months before traveling to Taos, he wrote an essay called "The Subject". His beliefs about the nature of painting colored his attitude towards the West. By this time his interests had turned decidedly away from realism and towards Modernism as he maintained that painting was first of all an abstract arrangement of formal relationships: "The elements that go to make the picture on your panel are — SHAPE, COLOR, and the SIZE of the colored shapes in relation to one another and to the size of the panel."[28] He added, "I want the picture to be simple." In this quest for pictorial simplicity and directness he declared "The subject must be visualized as a colored shape of a certain size. This represents the entire process of painting so far as I can see." Davis saw the process of modern art to be one of concentration, condensation and reduction. To be meaningful as a painting, the subject had to be treated in terms of its elemental component parts. It had to be contained within clear and definite boundaries. He identified Modern form with the limited, discrete, circumscribed shape.

This philosophy is antithetical to the experience of the New Mexico landscape. It is not surprising that he was unhappy there. His Taos paintings are distinguished by several pictorial devices that helped reduce the scale and depth of the land. These include translating the landscape into parallel planes that tie the image to the surface of the canvas and using vignettes. Adding painted frames delimits a scene, narrowing a vast expanse into a small perceptual field. The New Mexico landscape was too wide for Davis, too rambling and limitless.

Inspired by Synthetic Cubism, Davis observed that the "*unit* of drawing is an angular contrast" and asserted that "there must be *real* geometrical variety in the drawing."[29] As he saw it, an artist had to instill an artificial geometric order into reality to make it meaningful as a work of art. The problem with the New Mexico landscape was that it was by nature geometric in appearance. Davis was faced with a subject that seemed ready-made, already determined by an artificial geometric order. John Sloan had said, "I like to paint the landscape in the southwest because of the fine geometrical formations and the handsome color. Study of the desert forms, so severe and clear in that

Stuart Davis
New Mexico Gate. 1923
oil on linen
22 x 32 inches
Collection of Roswell Museum
and Art Center, gift of
Mr. and Mrs. Donald Winston and
Mr. and Mrs. Samuel H. Marshall

atmosphere, helped me to work out principles of plastic design, the low relief concept."[30] What delighted Sloan about the desert only disturbed Stuart Davis. The geometry he wanted to create on canvas was already there, a fundamental part of the landscape itself. Painting it seemed more like copying than invention. This is why he complained that the land consisted of "forms made to order, to imitate."

Davis represents a particular trend in American art of the 1920s, the tendency for stylization and abstraction, where abstraction took the form of a simplification of objects seen in nature. In the 1930s there developed a separate attitude toward abstraction with the promotion of nonobjective composition, abstractions that consist of pure arrangements of form that make little or only secondary reference to actual objects.

Marsden Hartley
Window, New Mexico. 1919
oil on canvas
37⅛ x 29 inches
Collection of Hirshhorn Museum and
Sculpture Garden, Smithsonian
Institution, gift of
Joseph H. Hirshhorn, 1966
Photography by Lee Stalsworth

Marsden Hartley
New Mexico Recollections #6. 1922
oil on canvas
26 x 36½ inches
The Harmsen Collection

MARSDEN HARTLEY ■ Hartley's *Amerika* series did not last long. After leaving Berlin in 1916, he returned to America. In June 1918 he moved to New Mexico where he remained until the end of 1919. He explained that he made the trip for two reasons, hoping to find an inexpensive place to live and to improve his health. He did not like Taos but for reasons very different from Stuart Davis's.

Hartley saw the majority of the white people there as provincial and small-minded, especially the recently formed Taos Society of Artists. Their strong presence in Taos prompted Hartley to call it "the stupidest place I ever fell into . . . a society of cheap artists from Chicago and New York."[31] This group consisted of representational artists, mostly illustrators, who settled in Taos and depicted the Indians in terms of current clichés. For Hartley their paintings pandered to common taste. It had no spiritual content and revealed nothing of significance about either the artist or his subjects. Instead of delving deeply into the true nature of the Native Americans, the Taos artists depicted them as stereotypes.

In opposition to the perceived opportunism of the Taos Society of Artists, Hartley called for what he termed "aesthetic sincerity" which he defined as a willingness to address personal feelings. "Regarding the southwest of America specifically, it is gratifying that there are tendencies appearing which at least indicate on the part of certain painters, the desire to think and feel in terms of the subject, and of themselves."[32] He reasoned that "genuineness, authenticity, are qualities we demand of ancient objects and pictures in the museums of art anywhere in the world. . . . We have a right then to expect the same characteristics in a modern painting, authenticity of emotion."[33] Finding authenticity in the directness of the Southwest landscape, Hartley wrote that "any one of these beautiful arroyos and canyons is a living example of the splendor of the ages . . . and I am bewitched by their magnificence and their austerity; as for the colour, it is of course the only place in America where true color exists, excepting the short autumnal season in New England."[34]

At the same time, Hartley admitted his difficulty in addressing such a magnificent subject. He wrote, "this country is very beautiful and also difficult. It needs a Courbet with a Renoir eye."[35] He was referring to Courbet's skillful representation of physical textures and Renoir's ability to capture subtle shifts in light. Hartley never attained this ideal balance. In fact, he found that he could paint powerful images of the New Mexico desert only after he had left the region. Working from his imagination in New York and later in Berlin, he produced his *New Mexico Recollections* series, which includes some of the most expressive evocations of the desert ever painted.

His *Window, New Mexico* of 1919, a view of the desert through a window, probably was painted while he was still in New Mexico. Although he had not arrived at a satisfactory solution to pure landscape in Taos, he was comfortable painting still lifes of inanimate objects. This painting is similar in composition to a series of works he executed in Taos of Mexican *santos*, devotional Christian images that reflect pre-Christian object worship. Hartley liked the motif of a view through a window because it offered him the opportunity to create a highly structured, symmetrical image. The composition, depicting a flower pot resting upon a table with a desert mountain outside, is as formal and hieratic as the paintings from his *Amerika* series. By treating the image in terms of simple abstract shapes, Hartley gave form to his sincere vision.

Hartley left New Mexico in November 1919 and returned to New York, not wanting to face another cold winter at Taos's high elevation. He began a series of large format paintings of the desert, painted from memory. In this first group of *New Mexico Recollections* he translated the curving hills of the desert into looming, ponderous masses of great power. In November 1921 he went back to Berlin where he embarked on a second phase of desert images. *New Mexico Recollections #6* of 1922-23 is dominated by squarish, blocky forms. In this airless painting, conceived in terms of mass and plane, space is compressed. The harsh angular faceting of the craggy rocks is exaggerated into an image of great expressive power. Working from memory, Hartley was able to create a pictorial structure as solid and massive as the landscape itself. He wrote that American artists, "will have to realize that the country of the southwest is essentially a sculptural country, and that there is no place in the world where the architecture has been made to fit the scene more perfectly, no instance of where nature itself has so completely dictated the human adaptation. The houses are like the earth because they are made of the earth, and are beautiful as the earth is beautiful."[36]

RAYMOND JONSON ■ Another artist who became interested in the natural rhythms of the earth was Raymond Jonson. In the 1920s his concern with the grandeur of nature continued but he shifted his attention from the transcendental glow of light to the internal rhythms within the earth. He visited New Mexico in 1922 and decided to move there permanently in 1924. In the 1920s he focused his attention on finding pictorial equivalents for the impressive features of the land.

His *Pueblo Series, Acoma* of 1927 belongs to a series of three oils depicting the pueblos of Taos, Acoma and Jemez, executed between 1926 and 1928. Jonson emphasized the blockiness of the rock formations in a way similar to Hartley. The important difference between the two is that Hartley's art is an expressive realism that remains wedded to visual experience, Jonson's is more mystical. It seeks to find the geometric and, by extension, metaphysical order behind reality. He wrote in his diary in 1923, that "in expressing my idea of this country

I struggled especially to obtain a unity—a unity of all the means used [such] as form, design, color, rhythm, and line."[37] This desire to find the essential order behind appearance is what will lead him in the direction of pure abstractions in the 1930s. It is not by coincidence that he chose to depict pueblos, sites of an ancient culture that represent centuries of knowledge and appear to embody a metaphysical truth in their accumulated wisdom.

These three Pueblo paintings are closely related to a larger group of eleven large oil paintings called the *Earth Rhythm Series*, painted between 1923 and 1927. They reflect a growing interest in abstracting from natural phenomena, a process Jonson referred to as "design," and believed to be a basic "unifying principle" of all art and life. Of one of these works, similar in style to *Pueblo Series, Acoma*, he said, "the forms are easily recognized as earth erosion, sky, etc. But all the means used are adapted to a certain concept of color, movement, space, light and dark and with at least some semblance of order result-

ing from the way in which it is planned."[38]

This quest for order is somewhat classical in spirit even though the subject is decidedly nonclassical. Sharyn Udall noted that in the Pueblo series "the viewer is reminded by repeated vertical striations of the fluting on a classical column and subjected on the other hand to a distinctly nonclassical surface turbulence in the painting."[39] His contact with the land prompted an examination of the geometric order that underlies all of reality. He wrote, "I have recently come in touch with an idea of a geometrical framework or governing system of direction in composition . . . I have always felt that there should be a governing sense of arrangement, that is, that a composition usually should have order and most often a simple basic motive of spaces, and interesting variety of shapes and spaces, a balance of line direction."[40] He added that "geometry has always been connected in some way or other with art. But I see in this something concrete." In the semi-geometric faceting of the angular Southwest landscape he was able

Raymond Jonson
Pueblo, Acoma. 1927
oil on canvas
37 x 44 inches
The Harmsen Collection

30

to perceive the reality of geometry as the metaphysical basis of the physical world.

CONRAD BUFF ■ Conrad Buff (1886-1976) was born in Switzerland where he studied lacemaking before immigrating to America in 1905. Because he did not speak English well he sought out communities of Swiss immigrants, settling in Wisconsin where he found work on a dairy farm. He moved across the country, taking odd jobs along the way before finally arriving in Los Angeles in 1907. Determined to become a fine artist, he began his career painting landscapes throughout Southern California and the Southwest.

His *Jawbone Canyon,* circa 1925, is painted in the modified pointillist style that characterizes Buff's work. Perhaps this technique was inspired by his training as a lacemaker; in his paintings fine threads of color are interlocked to create a vibrant surface of shimmering color. This composition is particularly striking, with strong contrasts between angular and rounded rocks, warm yellows and cool blues, and bright light and dark shadow.

Buff's organization of space is carefully studied with dramatic divisions between foreground, middle ground and distance. The position of forms is reinforced by distinct areas of sunlight and shadow. The overall effect is similar to that of a stage set where depth is developed artificially through the calculated use of depth cues. There is a similarity between the careful structure of Buff's landscapes and that of Maynard Dixon's. The two enjoyed a long and rewarding friendship.

Conrad Buff
Jawbone Canyon. ca. 1925
oil on canvas
47 x 66 inches
The Buck Collection, Laguna Hills,
California

31

Maynard Dixon
Cloud World. 1925
oil on canvas
34 x 62 inches
Collection of Arizona West Galleries,
Scottsdale
Photography by Bill McLemore

MAYNARD DIXON ■ Dixon's interest in the people and land of the Southwest continued the rest of his life. In 1920 he married photographer Dorothea Lange, who shared his enthusiasms. Through the 1920s they often traveled across the Southwest, working together to record the hidden beauty of the arid land. In formal terms Dixon continued working with juxtapositions of flat two-dimensional pattern and tangible three-dimensional form. *Cloud World* of 1925, one of his most impressive paintings, is built upon an abstract semigeometric pattern underlying the naturalistic forms. The bottom edge of the clouds form a definite plane that recedes dramatically into the distance. In this way Dixon emphasized the abstract structure behind reality, making all of nature, even intangible clouds, seem eternal.

John Marin
New Mexico, Near Taos. 1929
watercolor
14¹/₁₆ x 21¹/₁₆ inches
Collection of Los Angeles County
Museum of Art, The Mira T. Hershey
Memorial Collection

JOHN MARIN ■ When John Marin (1870-1953) first arrived in Taos in 1929 he was one of America's premier modernists. Born in Rutherford, New Jersey, he studied at the Pennsylvania Academy of Fine Arts from 1899 to 1901 and then at the Art Students League in New York. On a trip abroad, which lasted from 1905 to 1910, he pursued printmaking in earnest and produced etchings in Whistler's tonalist style. His art changed abruptly in 1912 when he began to paint cityscapes in an energetic style based loosely on Cubism. He utilized a Cubist fracturing of form, not to analyze a static object, but to express the dynamics of urban life. In these new paintings the forms of buildings shatter and shift with the pulse of a living organism. When he turned to painting landscapes in this new style, these lines of force represented the antagonisms between rock and tree, earth and sky. Marin became one of the central figures of the Stieglitz circle and one of the greatest American painters of the twentieth century.

In the spring of 1929 John Marin was encouraged by Mabel Dodge Luhan, a patron of the arts, and Georgia O'Keeffe to spend the summer in New Mexico. He accepted their suggestion, longing for a quieter pace. He spent two summers at Taos, from June to October 1929

and from mid-June to mid-September 1930, and painted nearly one hundred watercolors.

When he first arrived in Taos, Marin wrote to Stieglitz telling him of his plans for painting in the desert:

> I bought the Ford loaned to me by Mabel Luhan so that I am set — to set out in various directions in this huge layout, to be expressed by me on pieces of paper, with no effort whatsoever. . . . All my pictures this year will have labels tagged to them, as, this is so and so mountain I'd have you understand, and is, I'd have you understand, so and so feet high. Yes, all labeled with explanations like a *map*. That's the only way I know of, of getting by with this *dogoned* country.[41]

Marin's initial perceptions of the Southwest were similar to Stuart Davis's. Both perceived the land to be too overwhelming in its profusion of detail. Marin, however, understood the challenge and realized that he had to find a means to make sense out of the confusing countryside. His humorous suggestion of creating his work based on a map contains a kernel of truth; it reveals his awareness that he needed a

33

system to deal with the unique landscape of New Mexico.

As seen in *New Mexico, Near Taos* of 1929, one solution was to rely on a few simple shapes that he used throughout the composition. The ground is depicted with parallel, horizontal bands of color. The mountains consist of triangular forms set one against the other. Even the sky takes on this basic rhythm as dark bands of clouds radiate in a pyramid from a central point. Marin welcomed the largeness of the landscape for it offered him a means to resolve nature into large dominant masses. The grandeur of the mountains invited respect. This was very important to Marin. He once wrote to Stieglitz that his hosts Tony and Mabel Dodge Luhan were constantly repairing their adobe buildings because of weather damage. They were "everywhere repairing the damage wrought by the elements. So that here you are forced to *respect* the elements. A great thing to be forced to *respect* something."[42] The Southwest landscape offered a suitable subject because it was something the artist could respect.

Marin also utilized forms derived from Native American art to lend order and substance to his work. In *Taos Indian Rabbit Hunt* of 1929 he used the diagonal diamond shapes and step motifs borrowed from blanket patterns. Marin witnessed such a hunt personally and wrote:

> The Indians had a rabbit hunt a couple of weeks ago. *Yours truly* stumbled luckily right into the midst of it, almost taken for a rabbit. The rest of the bunch here not so fortunate. So that I am at work on an Indian rabbit hunt picture of immense proportions, size 8 inches by 10 inches. I am doing a bunch of sketches. The rest are painting pictures. As usual my studio is the big out-of-doors, in the car when too hot.[43]

Besides hunts, he also attended an Indian ceremonial dance that he found to be a remarkably moving experience:

> A big Indian dance I attended—I feel my greatest human Experience—the barbaric splendor of it was magnificent.
> The movements within movements are swell—and it kept up for hours.
> I drove an hundred miles to this dance—but that's nothing here—the country is so damn big—So that if you succeed in Expressing a little—one ought to be satisfied and proceed to pat oneself.[44]

He was greatly moved by the spectacle and in another letter he wrote:

> Certain passages in the dance itself are so beautiful that to produce a something having seen it—becomes well nigh worthless—it's like grafting on to perfection—it's like rewriting *Bach*.
> To out brilliance the diamond—to out red the ruby. But

man will always continue it seems to try and do just that.[45]

The paintings that Marin produced during his two summers in Taos are based on picturesque subjects. Although he had developed a distinctly modern vocabulary his subjects remained quite conventional. This is seen in his paintings of Indian ceremonies as well as in watercolors such as *Near Taos #5*. This work depicts the rather typical Taos subject of a horse and cart. Marin was always deeply moved by the simple things that create a sense of place. He enjoyed Taos because this town possessed a unique character. He once said:

> HERE is worth two of New Mexico out in the Bush—Maybe a settled location is best—Maybe a fellow ought to be chained to a settled location—maybe the Indians had it solved but alas they've been forced to part with their chains and are now a roaming *a la* white man—there are a bunch of them now down the road—a singing—I guess in this instance they are hired to sing and jump for the white man.[46]

GEORGIA O'KEEFFE ■ O'Keeffe's great contribution to twentieth-century images of the West lay in her tendency to focus on the sensuous qualities of the desert, which most people assume to be dry, harsh and ugly. She broke with the rigid angularity that dominated the paintings of artists such as Marsden Hartley, John Marin and Raymond Jonson. Rather than conceive of the landscape in terms of hard planes in opposition to one another, she focused on aspects of the desert that are soft and yielding.

Leaving Texas in 1918, O'Keeffe did not return to the West until the spring of 1929 when, along with John Marin, she spent four months in Taos, as a guest of Mabel Dodge Luhan. During the 1920s she had divided her time between New York City and Lake George in upstate New York with her husband Alfred Stieglitz, but she never felt comfortable in this environment. After arriving in Taos she wrote, "you know I never feel at home in the East like I do out here—and finally feeling in the right place again—I feel like myself—and I like it. . . . One perfect day after another—everyone going like mad after something—even if it is only sitting in the sun."[47]

The solitude of the area appealed to her greatly. Because the land was largely unpopulated it was easy to be alone. In isolation O'Keeffe was able to focus on her own thoughts. She needed this opportunity to examine her personal perceptions because they differed from the norm. For example, she found beauty in the barren rolling hills that most people thought to be ugly and recorded her perceptions in paintings such as *The Red Hills Beyond Abiquiu* of 1930. About these hills, she said:

> A red hill doesn't touch everyone's heart as it touches mine and I suppose there is no reason why it should. The red hill is a

John Marin
Taos Indian Rabbit Hunt. 1929
watercolor
21 x 16 inches
Collection of University of Maine at
Machias Art Galleries, Machias, Maine

John Marin
Near Taos #5.
watercolor
13¼ x 17½ inches
The Harmsen Collection

35

Georgia O'Keeffe
The Red Hills Beyond Abiquiu. 1930
oil on canvas
30 x 36 inches
Collection of the Eiteljorg Museum of
American Indian and Western Art, gift
of Harrison Eiteljorg

piece of the badlands where even the grass is gone. Badlands
roll away outside my door—hill after hill—red hills of apparently the same sort of earth that you mix with oil to make
paint. All the earth colors of the painter's palette are out there
in the many miles of badlands. The light Naples yellow
through the ochres—orange and red and purple earth—even
the soft earth greens. You have no associations with those hills
—our waste land—I think our most beautiful country.[48]

She was able to see promise in land that was rejected as wasteland and
perceive the great variety of color in ground that is often thought of as
grey and colorless.

The subject of *The Red Hills Beyond Abiquiu* is essentially one group
of hills, seen close up and set in the center of the painting. O'Keeffe
animates this large land mass by breaking it up into smaller shapes
divided by ravines. The ravine is the one recurring element in
O'Keeffe's desert paintings and is probably as important as her paintings of bones. A ravine is not the same as an arroyo, which is a dry
streambed with sharp, almost rectilinear, sides, formed by flowing

waters that cut into the hard earth. The ravine is a fold in the earth. Even when they become accentuated by years of erosion, they still have the character of a gentle crease in the ground that is distinct but not aggressive. In fact, they are the desert counterpart to the folds that O'Keeffe admired so much in her flower paintings, which constituted her signature image before arriving in the desert.

Like the petals of a flower, the ravine is a symbol of protection and comfort. Its sides enclose and offer shelter from the elements. The ravine is life-giving; serving as a conduit of water from runoff and springs, it provides a haven for plants that would not be able to grow elsewhere on the open desert. O'Keeffe emphasized the irregular undulations that move through the desert. They gently but unexpectedly twist and turn, like living things. O'Keeffe appreciated and sought to capture the natural liveliness in the physical processes that formed the ravine.

Artist Roger Brown, also included in this exhibition, was strongly influenced by the way O'Keeffe depicted folds. Perceptively noting their unique quality, he said, "she paints that reverse kind of way, as if the forms are reversed. Instead of the light hitting so the high surfaces are lit, it's the opposite. All the high surfaces become darker and behind is where the light comes out."[49] He meant that in normal representational painting the crevice is the darkest part because light is obstructed and cannot shine directly upon it. In O'Keeffe's work the deepest part of the fold is full of light. In this way she was able to imbue the still ground with a sense of vitality.

O'Keeffe made the arid rolling hills appear inviting rather than forbidding. They even seem pretty and she acknowledged that this was an important aspect of her art. When most other artists wanted their work to be serious and not merely pretty, O'Keeffe took the opposite view. "I'm one of the few artists, maybe the only one today, who is willing to talk about my work as pretty. I don't mind it being pretty. I think it's a shame to discard this word; maybe if we work on it hard enough we can make it fashionable again."[50] For O'Keeffe, pretty had a meaning greater than just the pleasant. Because she felt her perceptions to be so individual and so private, finding something to be pretty reassured her of the correctness of her own intuitions. She didn't use the word "beautiful," which has a long tradition in aesthetics and art theory through the centuries. "Pretty" is less determinate, more personal and more fickle.

As seen in her depictions of the desert hills, O'Keeffe was drawn to signs of life in inanimate matter. It is not surprising that she should be interested in Indian kachina dolls for a similar reason. In 1931 she mentioned in a letter a "small kachina — Indian doll — with the funny flat feather on its head and its eyes popping out — it has a curious kind of live stillness."[51] The kachina seemed to possess the energy of a real person. The strength of O'Keeffe's art lay in her ability to make the small seem monumental and the grandiose seem intimate. The kachina is a tailor-made subject since this figure is part of the Hopi religion but is also a toy. It is both larger and smaller than life. It is spiritual but also playful.

In 1931 O'Keeffe painted a full-length kachina doll given to her by photographer Paul Strand. She wrote, "I had the doll and made up what's behind it in the painting. It was the first painting of an Indian doll that I made. It was the first doll I owned."[52] Apparently O'Keeffe executed about seven paintings based on kachinas between the years 1931 and 1946, including *Katchina* of 1936.

O'Keeffe was drawn to bones for the same reason she was drawn to kachina dolls — both were unliving surrogates for living things. O'Keeffe felt most comfortable with the organic when it was controlled or removed in some way. In her flower paintings it was reduced to a semi-geometric patterning. In the bones she found in the desert, she sensed a residue of life that she preferred to living things and explained:

> I have wanted to paint the desert and I haven't known how. . . . So I brought home the bleached bones as my symbols of the desert. . . . To me they are strangely more living than the animals walking around — hair, eyes and all with their tails switching. The bones seem to cut sharply to the center of something that is keenly alive on the desert even tho' it is vast and empty and untouchable — and knows no kindness with all its beauty.[53]

Bones are lifeless matter that retain a sense of the life they once possessed. Bones are organic. Living bone is the core and center of every vertebrate animal. Why did she feel more comfortable with bones? Why did she feel such a strong affinity with them? One possible answer is that they did not provide a resistance to will as do living things. She seemed to disapprove of live "animals walking around — hair, eyes and all with their tails switching," probably because they

37

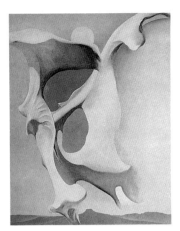

Georgia O'Keeffe
Pelvis with Moon. 1943
oil on canvas
30 x 24 inches
Collection of the Norton Gallery of
Art, West Palm Beach, Florida

were a potential nuisance. Bones never create problems, never assert their will.

She was particularly intrigued by the image of a large pelvis bone juxtaposed against the blue sky, as in *Pelvis With Moon* of 1943. She explained, "you see, I knew for a long time I was going to paint those bones. I had a whole pile of them in the patio waiting to be painted, and then one day I just happened to hold one up — and there was the sky through the hole. That was enough to start me."[54] The image of the pelvis against the blue seems to combine the bone with the sky. The limitations of the animal organism disappear in a transcendental fusion of the bone and the beyond.

O'Keeffe's great interest in bones cannot be considered apart from her lack of interest in other people. She did not like people in general. Once she asserted that "most of us are not even respectable warts on the face of the earth."[55] When she first arrived in Texas her initial impression was that "it's a pity to disfigure such wonderful country with people of any kind."[56] O'Keeffe felt that the prime beauty of the desert was that it allowed distance from other humans. After her second summer in New Mexico she wrote:

> One thing that gets me about . . . the Taos country — it is so beautiful — and so poisonous — the only way to live in it is to strictly mind your own business — your own . . . your own pleasures — and use your ears as little as possible — and keep the proportion of what one sees as it is in nature — much country — desert and mountain — and relatively keep the human being as about the size of a pin point — That was my feeling — is my feeling about my summer — most of the human side of it isn't worth thinking about — and as one chooses between the country and the human being the country becomes much more wonderful.[57]

O'Keeffe's rejection of people as bothersome was necessary for her to concentrate on her unique vision of the world. She saw things differently than other individuals in that she often focused on intangible perceptions that had no easily communicable form. This was the abstract element in her art. She thought of it as real since it was an aspect of her own emotional life. She once said, "it is surprising to me to see how many people separate the objective from the abstract. . . . The abstraction is often the most definite form for the intangible thing in myself that I can only clarify in paint."[58]

O'Keeffe considered her private perceptions of something to be real yet "intangible" because they did not adhere to any preconceived form. Her view of reality differed from that of most people. Abstraction allowed her to picture the world the way she saw it — simple and matter-of-fact, stripped of sentimentality and hypocrisy. The unadorned landscape of the West allowed her to realize this vision.

THE IDEAL OF ABSTRACT ART

JAY VAN EVEREN ■ There is not much known about Jay Van Everen (1875-1947), an artist and illustrator who produced abstract paintings and designs in a style strongly influenced by Art Deco. He worked throughout his life as a designer and book illustrator, and executed decorative mosaics for the New York City subway around the turn of the century. A turning point in his life occurred in 1917 when he met James Daugherty, who, from 1916 through the early 1920s, painted in a colorful abstract style based on Synchromism. Van Everen's abstractions of the 1920s display a heightened sense of decorative design. His paintings are dense, consisting of tightly knit shapes that vary from mathematically perfect geometric forms to freeform curves.

Amerindian Theme and *Untitled*, two compositions based on Native American subjects, are undated. Their streamlined stylization of naturalistic motifs is in the spirit of Art Deco, which had a pervasive effect upon the arts through the 1920s and 1930s. Van Everen painted them from photographs and did not visit the Southwest himself until the mid-1930s, when he was in his sixties. Because he did not have direct contact with his subject he was forced to invent many aspects of his images. This explains the high degree of abstract patterning and the arbitrary color. His eccentric coloration, whimsically combining bold hues with paler tones with no apparent rhyme or reason, is used to enliven the image. It developed from the artist's imagination and not from the actual hues of the Native American art he was copying. In these works, Van Everen was interested foremost in decorative design. He utilized the Indian subject not for any profound statement of content but as secondary to his concern for a well-designed, engaging image.

GEORGE L.K. MORRIS ■ George L.K. Morris (1905-1975) was a founding member of the American Abstract Artists, an association started in 1936 to promote the ideals of abstract art in this country. A cultured, well-educated individual, he enjoyed the financial means to pursue a style of art that was not successful commercially. Born in New York, he studied at Groton and Yale University. After graduation in 1928 he studied art with John Sloan and Kenneth Hayes Miller, two artists who practiced a form of socially conscious urban realism. A great change occurred when he traveled to Paris to study at the Académie Moderne with Fernand Léger and Amédée Ozenfant, both practitioners of Purism, a semiabstract style that sought an almost architectural exactitude. He returned in 1933 to New York where he became an active proselytizer for modern art.

Morris's involvement with Indian themes dates back to 1929 when he worked in a representational manner. *Indians Fighting #1* of 1934 is painted in a style reminiscent of Léger. The figures are conceived and

Jay Van Everen
Amerindian Theme. n.d.
oil and lacquer on masonite
46¼ x 45¾ inches
Collection of The Montclair Art
Museum, gift of Mr. and
Mrs. Rick Mielke
Photography by Steven Kasher

Jay Van Everen
Untitled. n.d.
oil and lacquer on masonite
26¼ x 37¼ inches
Collection of The Montclair Art
Museum, gift of Mr. and
Mrs. Rick Mielke

George L. K. Morris
Indians Fighting #1. 1934
oil on canvas
26⅛ x 22 inches
Collection of
Mr. and Mrs. Alan J. Pomerantz

rendered using pure formal elements. Only small sections of broken, curving outline define the two Indians; more emphasis seems to be placed on the floating, straight-edged areas of color. The reference to the subject of the painting is secondary to this ideal of clear, formal relationships. Morris once explained that his generation had "groped its way along and reached abstraction as a logical necessity."[59] They had come to understand how various modern movements since Cubism sought to do away with the excess baggage of representation and arrive at the essential structure behind reality. The members of the American Abstract Artists group conceived of the picture as an independent object that conformed to its own internal logic. They argued that since a painting is a flat arrangement of colors and lines, everything on it has to be conceived solely in these terms.

Morris's interest in Indians remained with him at least through 1951, and he returned to the theme often. For him the Native American represented an essential part of the culture of this country. Through the 1930s he became more and more interested in the inter-

nal dynamics of pictorial composition. He typically developed paintings such as *Indians Hunting #4* of 1934 in a series of sketches based on the visual force of abstract form. This painting exists in another version from the same year, *Indians Hunting #2*, which is the same essential composition only much simpler.[60] If these canvases were painted in the order of their numbering, we can see how he began with a simple pictorial idea and made it more complicated in the later work.

Morris did not develop his compositions in an arbitrary manner but tried to find the specific forms that best fit the general composition. He wrote that the artist is like a conductor in orchestrating the materials of his craft. "He will find that certain shapes as soon as they are set down *require* other particular shapes for the maintenance of specific positions, and similarly certain colors require certain other shades."[61] Morris focused his attention on such abstract pictorial considerations as the weight of different areas, the position of fixed mass, the movement between independent parts, the tactile sensation of texture and the quality of color.

George L.K. Morris
Indians Hunting #4. 1935
oil on canvas
35 x 40 inches
Collection of University of New Mexico
Art Museum, purchase through a grant
from the National Endowment for the
Arts with matching funds from the
Friends of Art
Photography by Damian Andrus

Raymond Jonson
Southwest Arrangement. 1933
oil on canvas
45 x 20 inches
Collection of Jonson Gallery of the
University Art Museum, University of
New Mexico, Albuquerque
Photography by John Waggaman

Phil Dike
Copper. 1935-36
oil on canvas
38 x 46¼ inches
Collection of Phoenix Art Museum,
purchase with funds provided by
Western Art Associates

This process may seem to be exclusively one of maintaining pictorial harmony, but this is an oversimplification because as Morris noted, "in some places the artist will purposely include two forces that jar, or he will keep two forces apart that seem to draw together because he wants that particular type of activity."[62] In both *Indians Hunting #4* and *Indians Fighting #1* the subject helped underscore this feeling of strife or opposition. These two activities emphasize antagonism. Morris found in the Indian a subject of primal force conducive to the pictorial concerns he had at the time.

RAYMOND JONSON ■ By the late 1920s Raymond Jonson was experimenting with pure abstraction. But instead of basing his abstractions on the internal logic of autonomous pictorial form as did Morris, Jonson looked to the process behind nature. From the time of his *Earth Rhythm* series, he began to examine carefully the essential processes underlying reality and discovered them to be abstract principles.

Southwest Arrangement of 1933 is an abstraction based on traditional Southwest Indian designs. But Jonson did not just reproduce patterns. The use of triangular shapes, which symbolize mountains, and transparent "rays" of light at the top, which symbolize sunlight, indicate that the artist is concerned foremost with relating the decorative design to the landscape and, by extension, to the world at large. He wanted to reveal the relevance or truth of these traditional Native American symbols to the processes of nature.

When he painted *Southwest Arrangement* he was also involved with a number of series based on organic growth and the cycles of nature, including the *Seasons* and *Growth Variant* series. From these works he realized a new conception of art, one that had nothing to do with design or representation. He said, "Art is not the business of making pictures. That is simple. It is rather something like creating a new rhythm, a new reality. It becomes a struggle to express some universal law as we find it in life and nature."[63] *Southwest Arrangement* reflects his awareness that the Native peoples had already discovered such a universal law.

42

The Great Depression

In the 1930s the Great Depression left its mark on all of the United States. Even the West, a land of great promise and hope, bore the scars of economic devastation. The economy of the West was particularly fragile since it was in a process of growth and had not developed the diversity that might help it through the bad times. The major markets and manufacturing centers were still in the East. Many individuals in the West earned a living providing the raw materials for factories in the Midwest. When these factories began to falter, the demand for the goods of the West disappeared. The West was also ravaged by the Dust Bowl, caused when speculators plowed millions of acres of marginal farmland that later turned to dust with the drought of the 1930s.

PHIL DIKE ■ A particularly powerful representation of this despair within a land of great natural beauty is Phil Dike's *Copper* of 1935-36. Its form embodies the upheavals that shook all of American society. A native of Los Angeles, Dike (1906-1990) studied at the Chouinard School of Art before traveling through Europe. After his return to Los Angeles in 1931 he visited Arizona, where he painted a series on copper mining.

While in Arizona, Dike became intrigued by its desolate mining towns. *Copper* represents the Arizona mining town of Jerome which is precariously situated on the side of steep hills. The artist remembers being moved by the uncertainty of life there, remarking, "how it holds onto the side of the Verde Valley no one knows."[64] His general impression was one of awe for both the harshness of the land and the fortitude of those who lived and worked in so inhospitable a place.

43

With the onset of the Depression, the plight of impoverished workers across America appeared a worthy topic for painters to address. The artist, who usually labored without great financial reward, could easily sympathize with the lives of those individuals living on the margins of society. In Arizona, Dike was particularly taken with the moving relation that existed between the dramatic land and the poor workers. He wrote:

Man tackling the vastness of that country, digging and living among cliffs and crag-riddled mountains, is so tremendous as to scare the sense of reality into one! Man-made forms and nature's giant ones, with the contrasting elements of thunderstorms and sunsets, were setting the stage for a humble but excited painter. The Depression had also left its mark, as mines closed and towns seemed to shrink under the sun.[65]

As with many other artists of the 1920s and 1930s, Dike was instinctively drawn to the semigeometric shapes of the "cliffs and crag-riddled mountains." He differed from the pure landscapists in that he chose to include an anecdotal element and depict human beings struggling for existence with the landscape. Since the nineteenth century, artists painting in the West have pictured the white settlements, but they are usually shown as picturesque genre scenes. Here Dike portrays the desperate relationship between miners and the earth. People and buildings are shown small-scale, dwarfed and made insignificant by looming mountains that threaten to swallow them.

The unusually high vantage point allowed the artist to explore imaginative relationships between parts of the scene. Although the work is representational, Dike did not approach it in terms of a straightforward realism, merely copying what he saw before him. He exaggerated features and distorted elements to create a more expressive image. As he explained:

I might point out my attempt to build up my light and dark pattern to accept the mine plant; also the character and the unique position of the town, which might be a document pertaining to Arizona camps in general. I attempted to combine in one arrangement the lighting and atmospheric effects which to me suggested the many dramatic moods of this country.[66]

Phil Dike believed that one of the important factors of any work of art is "the idea or conception of the subject matter with a mind to its emotional drive or graphic fitness."[67] He referred to this as "staging." The choice of words is particularly apt since *Copper* is a highly staged painting. It is conceived in terms of an artificial spatial construction populated with miniature buildings and people, much as one would place figures within a diorama.

Although Dike's work is traditional in many ways, its use of staging is directly inspired by the modernist concept of constructing a

pictorial reality separate from that of the natural world. He said that in this series he consciously chose to use "overlapping circular, triangular and square forms" as the basis of his compositions. He acknowledged that "this might have been carried out more abstractly. However, I wished to use it as a foundation on which to pin the idea."[68] The hard-edged forms and geometric shapes were used to give substance to the image. But this formal treatment was also inspired by the land itself, "the sharpness of the delineation seemed characteristic of that country, as well as the use of secondary forms that are typical of the Southwest."[69]

LEW DAVIS ■ Since Lew Davis (1910-1979) was born in the copper-mining town of Jerome, Arizona, he knew first hand the experiences that Dike depicted. After enrolling at the National Academy of Design at the age of seventeen, he spent much of the late 1920s and early 1930s in the East. A WPA program provided the incentive for his return to Arizona. He recalled that "in 1935 the Treasury Department Section of Painting and Sculpture began a project that encouraged artists to go back from where they came. Under the plan one received a monthly salary and was obliged to send to Washington about four paintings a year to be used to decorate public buildings. . . . I established a studio in the unused sample rooms of the Jerome Hotel and went to work."[70] After marrying in 1937, he moved to Scottsdale but continued working on the *Jerome* series until about 1940.

Little Boy Lives in a Copper Camp of 1939 is one of the most moving images from the *Jerome* series and is probably Davis's most famous painting. In this work a sad-faced young boy sits pensively in a sparse interior. Through the window behind him one sees a barren landscape populated by the functional buildings and sheds used by mining companies. The poignancy of this image is heightened by the boy's complete detachment. He looks neither toward the viewer nor out onto the landscape; he knows that neither direction holds a future for him. He reacts not with desperation or despair but with solemn resignation to his situation.

The *Jerome* series belongs to the style known as Regionalism that dominated American painting during the Depression. The movement reacted against the influx of European modernist styles by advocating a return to realism and a concentration upon indigenous American themes, usually drawn from the farm life of America's heartland. In the hands of more programmatic Regionalists such as Thomas Hart Benton, this style was used to condemn urbanism by promoting a positive view of traditional American values, lauding such staples of rural life as the small town, the family farm and the local church. Lew Davis's painting is more somber and far less optimistic. Living in a depressed Arizona mining town, he experienced deprivation firsthand and could not romanticize the plight of the impoverished miners into

Lew Davis
Little Boy Lives in a Copper Camp.
1939
oil on masonite
29½ x 24½ inches
Collection of Phoenix Art Museum, gift
of IBM Corporation

a statement glorifying those who make their living from the land.

It has been noted that Davis's sources for the *Jerome* paintings included the Italian primitives of the Quattrocento such as Paolo Uccello and Piero della Francesca, as well as the contemporary murals of Diego Rivera. For the Italians of the early Renaissance, their imperfect, almost awkward, realism reflected lingering medieval tendencies in their art. Davis adopted their exaggerations and tentative use of perspective to imbue his subject with a harshness and crudeness that reflects the conditions of life there. He also took from Italian tempera panel painting the use of a green underpainting, which adds to the feeling of melancholy surrounding the figure in *Little Boy Lives in a Copper Camp*. From Diego Rivera he borrowed a renewed sense of humanism based on the individual, filling the painting with a large-scale figure that dominates the composition. The starkness of his image is also in debt to German Expressionism and in particular to the New Objectivity, a harsh, brutally honest realism that arose in that country in the 1920s.

45

Adolph Gottlieb
Untitled (Still Life with Watermelon —
Dry Cactus). 1938
oil on canvas
23⅞ x 30 inches
Collection of Adolph and Esther
Gottlieb Foundation, Inc.
Photography by Peter Muscato

46

EARLY ABSTRACT EXPRESSIONISM

ADOLPH GOTTLIEB ■ The Depression paintings of Phil Dike and Lew Davis were essentially realist works that incorporated various formal elements drawn from modern art. They represent a period of retrenchment in American painting when artists, by now familiar with modern art, began to utilize modernist elements within paintings that remained traditional in spirit. A new vision arose at this time in the work of a few artists who were to become known as the Abstract Expressionists in the late 1940s. In the 1930s painters such as Adolph Gottlieb, Jackson Pollock and Richard Pousette-Dart drew inspiration from non-European sources — in part from their experience of the land and people of the West — to develop a new conception of painting, one based not so much on copying the external world or abstracting from it but on the internal consciousness of the artist himself.

Adolph Gottlieb (1903-1974) was a native of New York whose experience in the Southwest inspired an important change in direction in his art. A member of a figurative expressionist group known as The Ten from 1935 to 1939, he spent the winter and spring of 1938 in Tucson, Arizona where he embarked on a series of still lifes. They would prove important in the development of his mature Abstract Expressionist style which used abstract symbols as evocative, iconic signs, much as in Native American petroglyphs.

Gottlieb remembers being impressed by the broad expanse of the desert around Tucson and said, "I think the emotional feeling I had on the desert was that it was like being at sea . . . In fact, when you're out on the desert, you see the horizon for 360 degrees . . . so that the desert is like the ocean in that sense."[71] Although he was impressed by the sensation of the desert, he was unable to paint it at first. After being in Arizona for about a month he wrote to a friend, "So far haven't been able to do anything with desert landscape, so gave it up for a while" and concentrated on sketches based on an earlier visit to Vermont.[72]

Perhaps the feeling of "being at sea" was too disorienting and did not offer a stable point of focus. He solved this problem by temporarily abandoning the vexing landscape and "painting still lifes — particularly chess men and gourds."[73] He soon abandoned the chessmen but continued painting the gourds and other crops popular in the Southwest. "We are still plugging away at still life — no more chessmen — garden vegetables for a change — and getting lots of work done."[74]

In paintings such as *Untitled (Still Life with Watermelon — Dry Cactus)* and *Symbols of the Desert*, both of 1938, Gottlieb concentrated on the curious examples of plant life found in the desert. The desert flora seemed particularly odd because it was living but seemed lifeless. The cactus, cholla and other native plants stood in stark contrast to the moist garden produce introduced and cultivated by white farmers. This opposition is most evident in *Untitled (Still Life with Watermelon — Dry Cactus)* where a watermelon, a plant containing an unusually high percentage of water, is juxtaposed with dry cactus.

Gottlieb used these contrasts to meditate on the very nature of life itself. How can something that appears to be dead actually be alive? Is the life within a dry cactus different from that of the watermelon plant? In a way Gottlieb's reaction to the dry debris of the desert is similar to Georgia O'Keeffe's. But Gottlieb's work is more searching, more concerned with fundamental metaphysical questions about the nature of existence. O'Keeffe celebrated the formal beauty of a pelvis juxtaposed against the sky. Gottlieb seemed to be interested foremost in grasping the inner nature of these strange objects. His interest in the metaphysical status of objects continued in his mature work where he explored the nature of different symbols and sign systems, contrasting those signs we readily understand and those that speak to us obliquely.

Gottlieb's still life arrangements are usually juxtaposed against a desert landscape. This solution allowed the artist to finally master the emptiness of the countryside around Tucson by including it as a subsidiary motif in the background. After four months in Arizona Gottlieb could write, "I think I've gotten the hang of landscape at last. I mean a way of approaching the subject. Never thought I was cut out to be a landscape painter but may be I'll be one yet."[75] *Symbols of the Desert*, a work painted immediately upon Gottlieb's return to New York, includes the desert as a view through a window. On the table in the foreground are signs of the desert beyond. Specimens of cactus are presented under glass next to two eggs (archetypal symbols of birth), again an obvious contrast of life and the lifeless.

The strange and unfamiliar life forms of the desert prompted Gottlieb to shift his attention from Social Realist themes prevalent among members of The Ten to metaphysical questions about the nature of life itself. At the same time, the bright raking light of the desert, which threw all forms into strong relief, served as a model for a new simplification in his art. He noted that friends felt the experience in the Arizona desert helped make his work more abstract. He recalled:

> After spending a year in Arizona around 1938, I came back to New York with a series of still lives. Everyone said my paintings had become very abstract. The thought had never occurred to me whether they were abstract or not abstract. I simply felt that the themes I found in the Southwest required a different approach from that I had used before.[76]

47

48

Adolph Gottlieb
Symbols and the Desert. 1938
oil on canvas
39¾ x 35⅞ inches
Collection of Adolph and Esther
Gottlieb Foundation, Inc.
Photography by Otto E. Nelson
Photography by Peter Muscato

Jackson Pollock
Camp With Oil Rig. 1930-33
oil on board
18 x 25¼ inches
Collection of
Mr. and Mrs. John W. Mecom, Jr.,
Houston
Photography by Norlene Tips

JACKSON POLLOCK ■ Jackson Pollock (1912-1956) is known for his mature drip paintings that introduced a radically new conception of pictorial space and of the artist's physical involvement with his work. Born in Cody, Wyoming to a poor migrant family, he was raised in a number of different locations in the West, mainly in Arizona and California. *Camp with Oil Rig* of 1930-33, an early work, was done while Pollock was studying at the Art Students League in New York under Thomas Hart Benton, America's best known regionalist painter. Benton had Pollock in his classes from 1931 to 1932 and the two remained friends until Pollock's tragic death. He said that the young man's achievement in these early works was to "have injected a mystic strain into the more generally prosaic characteristics of Regionalism."[77]

Camp with Oil Rig breaks with a matter-of-fact representation through its air of mystery. The scene is murky, stark and desolate. The somber, greenish cast contributes to a mood of desperation and alienation. Pollock rejected the blind optimism advanced by earlier artists and depicted the West as a failed promise. He grew up experiencing firsthand another side of the West, the West of economic impoverishment. His childhood experiences reflect this sense of dissolution and lack of direction. The good life that was once promised to all never materialized. The West had lost its aura as a land of riches and opportunity. Pollock's rebelliousness and psychological confusion can be traced to a lack of feeling of centeredness. He no longer knew his place; the emptiness of the land did nothing to provide deep spiritual meaning.

In the late 1930s Pollock turned to Native American art and produced works such as *Composition with Horse*, circa 1934-38. The horse was present in Pollock's regionalist work from the time of his study with Benton. Along with the bull, the horse became more and more meaningful to him in the late 1930s and early 1940s, influenced to a great extent by Picasso's masterpiece *Guernica* in which the bull and horse represented human aggression and helplessness, respectively. Pollock was drawn to these primal images of instinctive animal impulses but could not share Picasso's European and classical sources for these subjects. He sought a new, more personal meaning in these forms. He found it in Native American art where animals such as the horse, bison and eagle appear as totemic symbols of nature's elemental forces. Pollock told an interviewer:

> I have always been very impressed with the plastic qualities of American Indian art. The Indians have the true painter's approach in their capacity to get hold of appropriate images, and their understanding of what constitutes painterly subject matter. Their color is essentially Western, their vision has the basic universality of all real art. Some people find references to American Indian art and calligraphy in my pictures. That wasn't intentional; probably was the result of early memories and enthusiasms.[78]

In *Composition with Horse* the animal is depicted schematically in a manner similar to the pictorial art of the Plains Indians. But there are profound differences. Pollock's painting is dark, densely packed, brooding. His various marks are tentative, sometimes conflicting and give a general impression of confusion and uncertainty as the artist struggled to give pictorial form to his troubled thoughts. The Native American artist, on the other hand, worked with symbolic forms that were deeply embedded in the collective psyche of the tribe. He could create painterly images without hesitation or doubt because the symbols he used were an integral part of his culture. The Indian artist worked out of a communal, shared understanding of meanings and values that led to a clarity and certainty lacking in Pollock's work.

Composition with Horse is black, almost muddy in tone. The critic Clement Greenberg, one of the first to appreciate Pollock's talent, noted this tendency toward muddiness. Usually seen as a sign of an amateur inability to handle paint, muddiness was reevaluated positively by Greenberg, who declared that Pollock was "the first painter I know of to have got something positive from the muddiness of color that so profoundly characterizes a great deal of American painting."[79] For Greenberg muddiness was a consequence of Pollock's "ambition," particularly of his willingness to go beyond the bounds of his ability and take "orders he can't fill." But this muddiness does result from inner indecisiveness and is the antithesis of the Indian art that inspired this work. In Native American art each particular form and color held a separate spiritual meaning. Pollock lived within a culture that endowed matter with no comparably profound significance. Unable to appreciate the uniqueness of his medium as a natural substance, it became coarse, broad and muddy in his hands.

During the late 1930s and early 1940s Pollock frequented the ethnographic galleries of the American Museum of Natural History and the Museum of the American Indian in New York. He is known to have made return visits to the Museum of Modern Art's important exhibition *Indian Art of the United States* of 1941. He even clipped and saved a newspaper article on this exhibition that illustrated ten Native American masks. *Untitled*, a drawing of 1943 currently in the Montana Historical Society, features masklike faces, a motif that appeared in his work since 1938. For the Native American artist the mask represented a distinct spiritual being, an artificial but integrated personality. Pollock longed to experience a stable psyche and used Native American art in part to regress into the depth of his own mind to resolve inner psychological conflicts.

In *Untitled* of 1943 in the Whitney Museum of American Art, stick figures resemble the painted figures from ancient Southwest pottery and petroglyphs. Pollock was also interested in the function of shaman artists in Indian societies. These shamans painted as a form of empowerment by conjuring visions. The rapid calligraphic marks capture the mood of swirling, gesticulating figures seen in a visionary trance.

Totem Lesson I of 1944 marked a return to an interest in totemic themes and Native American art. There is a new fluidity in this work as the artist tried to free himself from a dependence upon literal references. The sole recognizable element is again a profile mask, highly stylized and reduced to essential features. In large paintings such as this Pollock began to explore the gestural freedom that would lead to his innovative drip paintings of 1948.

50

Jackson Pollock
Composition with Horse. ca. 1934-38
oil on panel
10½ x 20¾ inches
Courtesy of The Gerald Peters Gallery,
Santa Fe, New Mexico

Jackson Pollock
Untitled. 1943
ink and watercolor on paper
26 x 20½ inches
Collection of Montana Historical
Society, Poindexter Collection
Photography by John Smart

52

Jackson Pollock
Untitled. ca. 1939-42
india ink on paper
18 x 13⅞ inches
Collection of Whitney Museum of
American Art, New York, purchase with
funds from the Julia B. Engel Purchase
Fund and the Drawing Committee

Jackson Pollock
Totem Lesson I. 1944
oil on canvas
70 x 44 inches
Collection of Mr. and Mrs. Harry W.
Anderson, Atherton, California
Photography by M. Lee Fatherree

Byron Browne
Variations on Haida Masks. 1934
watercolor and india ink on paper
19¾ x 11¾ inches
Courtesy of Michael Rosenfeld Gallery,
New York

BYRON BROWNE ■ Byron Browne (1907-1961), like his contemporary George L.K. Morris, championed nonobjective painting in America during the crucial years of the 1930s and 1940s. Born in New York, he studied at the National Academy of Design from 1925 to 1928. After becoming acquainted with Cubism and abstract art soon after graduation, he turned his back on his earlier academic work. In the 1930s he explored a constructivist type of composition based on the intersection of hard-edged geometric planes. His *Variations on Haida Masks* of 1934 uses such a compositional structure to interpret the bold stylizations of this Northwest Coast Indian art form. Unlike many of his contemporaries who conceived of abstract form in terms of solid, opaque planes, Browne embraced the concept of transparency. As seen here, he exploited the natural transparency of watercolor to explore the interpenetration of planes through space. Browne's art never achieved the power seen in the work of many of his contemporaries, in part due to his tendency for stylization that seems too decorative. At the same time much of his semi-abstractions tended toward the literal. *Variations on Haida Masks* strikes a balance between the art of the Haida people and modern abstraction.

RICHARD POUSETTE-DART ■ Richard Pousette-Dart (born 1916) is known for his large spiritual abstractions, all-over fields of pulsating color that allude to cosmic phenomena. For many years his work has been considered to be only a variant of the Abstract Expressionist movement. However a recent reappraisal of spiritual abstraction has done much to revive his reputation. Today we are able to see his work on its own terms, as representing a highly personal vision of the human ability to find oneness with the universe. It is important to remember that he developed this vision under the influence of primitive art, and of Native American art in particular.

Raised in an artistic family, Pousette-Dart enrolled at Bard College in 1936 but soon left school to study art on his own. He became interested in primitive art and began to visit the American Museum of Natural History. He recalled that the semiabstract animal forms in *Untitled, Birds and Fish* of 1939 were directly inspired by examples of Northwest Coast American Indian art in their galleries. He probably had seen a mid-nineteenth-century painted mural from the Nootka tribe depicting a thunderbird carrying a killer whale, flanked by a lightning snake and a wolf. Pousette-Dart remade the image based on his personal experience with animals. As he noted, the birds derived from pigeons seen near his home, the fish from those seen at the Fulton Fish Market in lower Manhattan.[80] He did not simply copy the Nootka painting but used it as a means to explore his own feelings about birds and fish, which are symbols with deep meaning for many cultures, including our own. Joanne Kuebler found that the bird and fish are both "ancient religious symbols and ubiquitous American Indian motifs. The fish was a secret symbol for early Christians and is

Richard Pousette-Dart
Untitled, Birds and Fish. 1939
oil on linen
36¼ x 60 inches
Collection of
Mrs. Richard Pousette-Dart
Photograph courtesy of Indianapolis
Museum of Art

Nootka tribe
The Thunderbird with the Whale in his
Talons, Lightning Snake and Wolf.
Vancouver Island, ca. 1850
mural on wood
64¼ x 118 inches
Courtesy of Department of Library
Services, American Museum of
Natural History
Photography by H.S. Rice, Neg. No.
313186

Richard Pousette-Dart
Primordial Moment. 1939
oil on linen
36 x 48 inches
Collection of
Mrs. Richard Pousette-Dart
Photography by Stephen Kovacik,
courtesy of Indianapolis Museum of Art

56

a symbol of Christ; the bird motif is pervasive in Egyptian hiero-glyphics."[81] More generally these animals reflect a desire to be in touch with the elements. The bird is a creature of the air, the fish is one of the water.

Pousette-Dart once said that "the most exciting work of our time grows out of primitivism. Nobody has defined so clearly the principles of dealing with pure form. The primitives make architectural monuments from the head or figure or animal abstracted from their visual and life experience."[82] What was important was that in this process of abstracting elements from their life experience, the primitive artist imbued his forms with life itself. For Pousette-Dart, this was most evident in the edges of a form, which he referred to as "the trembling living edges of living definition." He explained that "the edge is inseparable from the substance within the form, but it is a created definition and not a mechanical definition. Primitive artists make edges that may be so severe that you can cut yourself on them, but their edges live because they are created edges, won through feeling."[83]

He explored this interest in "living edge" in paintings such as *Primordial Moment* of 1939 which is also based on Northwest Coast Indian art. One recurrent feature of this art is the unique combinations of form and outline into one entity that has been called formline. Formline arises when the line circumscribing shapes thickens so that it becomes a shape in itself. In this way the distinction between negative and positive space becomes blurred. Elements of representation become highly schematized symbols. Pousette-Dart liked this because line did not simply ring an object but functioned as a living aura encircling it.

He once said, "I felt close to the spirit of Indian art. My work came from some spirit or force in America not Europe."[84] What he admired was the pantheistic quality of Native American art and religion. This work was in harmony with his belief in the oneness of art and nature. "All art is abstract and all abstract art is of nature because we are of nature. . . . Art for me is unpredictable spontaneous kaleidoscopes of the imagination."[85]

In talking about his later abstractions, Pousette-Dart said that his paintings were made from "beaten earth and sky" and asserted that he didn't want "an earth that hasn't been danced on."[86] This is a very spiritual idea, in harmony with the nature-religions of Native Americans. He meant that the forms of his art were primal, taken from the same ground from which all life springs.

WILL BARNET ■ Another artist who showed a marked interest in Northwest Coast Indian art was Will Barnet (born 1911). In the late 1940s he was associated with a short-lived group of artists who centered around the magazine *Iconograph*. They were united by their com-

mon admiration for American Indian art, particularly for its strong sense of two-dimensional design. They referred to this quality as Indian Space because it was opposed to the illusionistic perspectival space practiced by European artists since the time of the Renaissance.

Will Barnet's *Self Portrait* of 1948-49 was based on the flat patterning of Native American art, especially that of the Northwest Coast. He wanted to arrive at the mystery behind these works of art and said, "I tried to get at the emotions behind it. . . . Why did they make these forms, these angles, these poems?"[87] He recalled that his "search in the late forties was to find forms that belonged to the pure matter of painting itself but which were equivalent to the substance and the forces that I felt in nature. I eliminated realistic space and substituted a painting space based purely on the rectangle: the vertical and horizontal expansion of forms."[88] He was drawn to the art of Northwest Coast peoples because the space was "all positive," by which he meant that instead of consisting of a main positive shape surrounded by subsidiary negative shapes, every form had its own pictorial integrity.

ART IN THE NORTHWEST

The majority of the artists discussed so far chose to paint in the Southwest, yet an important school of art was emerging in the Northwest in the 1930s and 1940s. The Northwest painters were fewer in number but they developed a powerful art influenced to a great extent by the unique qualities of the environment. Their work tends to be less concerned with the abstract planes and simple masses that occupied the attention of so many Southwest artists. The Northwest is a land of forests and water. The predominant colors are green and blue, not red and yellow. Many hills and dense green foliage prevent broad views across great open vistas. The artists working there developed a sensitivity to the soft atmospheric effects found in a moist environment. Many chose to concentrate on close-up views and some even turned to look inward, examining mystical states of mind.

KENNETH CALLAHAN ■ Kenneth Callahan (1905-1985) typifies the approach of the Northwest artists. Instead of seeing the landscape as arid and hostile to life as so many Southwest painters have done, he looked upon the land as a nurturing, generative force. A recurring theme in his art is the process by which people emerge from and return to the larger matrix of nature. Born in Spokane, Washington, Callahan studied at the University of Washington in Seattle. He worked as a seaman from 1926 to 1928 and then as a forest ranger before deciding to become a painter.

Kenneth Callahan's Northwest landscapes of the 1930s and 1940s are similar to those painted in the Southwest by artists such as Hartley in that they concentrated on rugged, rocky forms. But Callahan's vision was wider ranging than that of his contemporaries in the desert. His Northwest landscapes almost always include people. These figures add a scale to the landforms but a new content. The subject is no longer the land itself but, rather, how people relate to their world. In *Traincrew* of 1940 he shows the efforts of those who labor to move people and goods through the high mountains and dense forests. In the early 1930s, one of his favorite themes was that of workers who earn their living from the land, such as lumberjacks and sailors. He appreciated the degree to which the destinies of people and nature are intertwined and once described his subject matter as involving "humanity evolving in and out of nature."[89]

Where Cézanne was a great influence on the predominant style of the Southwest painters, Callahan looked to a very different model. He was greatly interested in the Italian Mannerist El Greco since the late 1920s.[90] El Greco's elongated, spiritualized forms and flickering unearthly light had more significance to an artist in the Northwest where nature assumed the form of tall forests and high peaks. The

uneven light in Callahan's landscapes reflects the constantly changing light and weather conditions of the Pacific Northwest.

AMBROSE PATTERSON ■ Ambrose Patterson (1877-1966) was born in Australia and studied art there and in Paris before settling in Seattle in 1918, where he played an important role in establishing the School of Painting and Sculpture at the University of Washington the following year. His *Rocky Landscape* of 1946 is similar in style to Callahan's work but shows more of an interest in Oriental art, particularly Chinese landscape painting. The west coast of the United States faces the Pacific Ocean and the Orient. In the Southwest, far from the sea, this connection has little meaning. In the Northwest, where many people depended upon the ocean for their livelihood, it was of great significance.

In Chinese landscapes the artist invites the viewer to enter his pictorial world. Instead of admiring the work of art from afar, the spectator is expected to take a visual stroll through the painted space, imagining himself crossing each path, climbing every hill. Patterson does something very similar in this painting, recognizing the need to establish a symbiotic relationship between the individual and the work of art. Instead of viewing the work with detachment, a person is asked to become part of the experience on the canvas. This emphasis on empathy is an important aspect of the Northwest School, particularly in the work of its most mystical painters, Morris Graves and Mark Tobey.

MORRIS GRAVES ■ Morris Graves (born 1910) one of the most intensely spiritual of the Northwest artists, is known for paintings of solitary birds in which these frail creatures symbolize the aspirations and suffering of all living things. Born in Oregon, he moved to Seattle in 1912 and lived in or near that city for the next fifty-five years. In 1928 he visited Asia while working on a merchant ship and returned twice in 1930. He completed high school in Texas but returned to Seattle in 1932, where he was on the WPA from 1936 to 1939. A deeply introspective person, Graves was always concerned with the fate of mankind. A pacifist who opposed the dehumanization of the mechanistic world, he embraced Eastern metaphysical systems including Zen.

Graves has spoken often of the importance of the "inner eye," a state of consciousness where the meditating artist "sees" the essential truth. He became interested in mandalas, which are concentric patterns, usually circles, that function as an aid in meditation. In 1943 and 1944 he painted a series based on pine trees. Graves believed that the pyramidal structure of these images related to a form of a Tibetan stupa, a vertical, three-dimensional mandala.[91] In *Joyous Young Pine*

Kenneth Callahan
Trail Crew. 1940
oil on canvas
20 x 22⅞ inches
Collection of Seattle Art Museum,
Eugene Fuller Memorial Collection
Photography by Susan Dirk

Ambrose Patterson
Rocky Landscape. 1946
oil on panel
22⅜ x 30 inches
Collection of Seattle Art Museum,
Eugene Fuller Memorial Collection
Photography by Susan Dirk

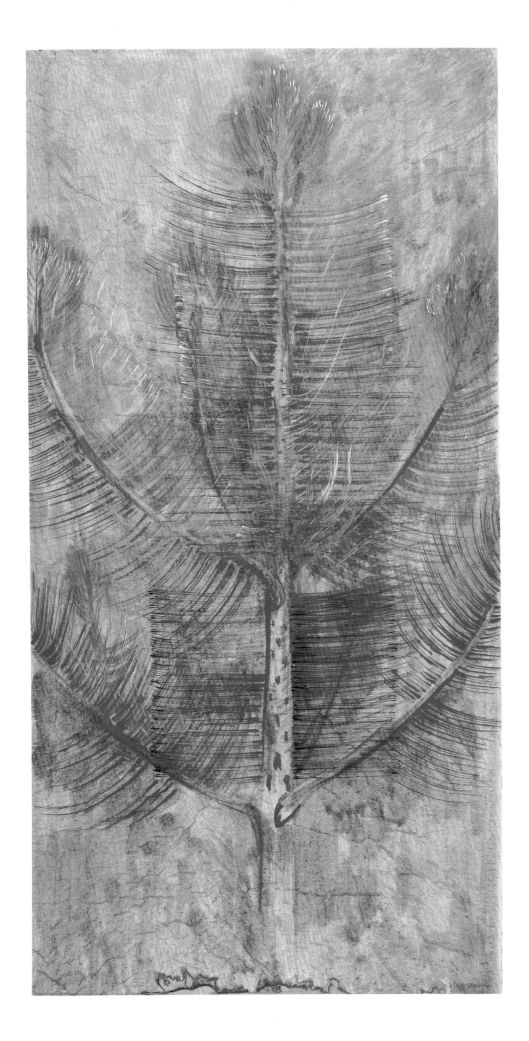

Morris Graves
Joyous Young Pine. n.d.
watercolor on paper pasted on
cardboard
52¾ x 27 inches
Collection of Santa Barbara Museum of
Art, gift of Wright S. Ludington

Mark Tobey
Mexican Ritual. 1931
oil on board
15¼ x 11¼ inches
Collection of Helen and Marshall Hatch
Photography by Paul Macapia

the branches correspond to different levels of the stupa, with each one signifying a higher state of consciousness. A spiritual progression to greater understanding is symbolized using the natural growth pattern of the young pine tree. Graves's empathy for living things found a parallel between the Northwest pine tree and eastern religion. Instead of seeing the tree in terms of its physical form, he perceives its spiritual essence.

MARK TOBEY ■

Mark Tobey (1890-1976) is known for his spiritualized abstractions based on fields of fine lines known as "white writing." Born in Wisconsin, he had little formal art training before leaving high school in 1909 to become a fashion illustrator, a career he pursued in New York and Chicago until 1917. Around 1918 he became an adherent of the Baha'i faith, which believes in the oneness of all peoples and in the essential unity of all religions. Baha'i maintains that the teachings of the world's major religions are all part of a divine progressive plan for the education and betterment of the human race. From 1922 to 1925 Tobey taught at the Cornish School of Allied Arts in Seattle, where he developed an interest in Northwest Coast American Indian art, Japanese woodcuts and in Oriental calligraphy, which he learned from a Chinese student at the University of Washington. In 1934 he traveled to the Orient where he furthered his knowledge of calligraphy and eastern religion.

Tobey's early efforts to transcend material reality included symbolic depictions of rituals. *Ritual* of 1936, also known as *Mexican Ritual*, depicts a group of people engaged in a ceremony. In keeping with the tenets of Baha'i, Tobey was interested in the general act of religious piety and not the specifics of any one ritual. To him this scene signified only the universal urge for religious gathering, not the particulars of the ceremony. In fact, Tobey misidentified the culture he depicted. Although the title indicates that this scene is in Mexico, the people portrayed are more likely Native Americans of the Southwest. Tobey did not care about national or tribal differences, for political boundaries mattered little when compared to the great faith binding all of humanity. He was taken with the religiosity of these people because it seemed natural and genuine, free from artifice. The pictorial forms used in *Ritual* are radically simplified in an intuitive way, conforming to no particular stylistic tendency. The shapes stand as signs or ciphers, almost archaic symbols, of the things they represent.

Tobey's interest in ties that bind all of humanity led to his involvement with paintings of cities and towns. His concern with cities began in the mid-1930s and recurred often through his formative years. Art historian Martha Kingsbury wrote that between 1935 and 1945, when Tobey discovered his mature style, he showed a great interest in two related subjects, cities and the Gothic. These "were

subjects in which he saw an overriding energy that dematerialized form; they provided a historical and a contemporary example of cultural constructs transcending the human and becoming universal."[92]

This fusion of the city with the Gothic is seen in his *Western Town* of 1944. In this work Tobey translates the persistent energy of a burgeoning western town into a quivering, grid-like structure of architectonic forms painted in his white calligraphic style known as "white writing". Tobey explained this unusual technique by saying, "People say I called my painting 'white writing.' I didn't. Somebody else did. I was interested in an idea—why couldn't structures be in white? Why did they always have to be in black? I painted them in white because I thought structures could be, should be light. What I was fundamentally interested in at the time was light."[93]

The use of white seems to dematerialize the buildings, making them seem otherworldly and transcendental. They are freed from the dirt, grime and facts of physical existence.

But for all its energy and inner light there is something calm, small and provincial about *Western Town*. The town is modest in scale; there is no mistaking it for a great metropolis. Tobey conveys this sense of smallness by leaving sizable margins around the structures, making it appear that the town has a definite end. In its intimate scale it seems to have a distinct character, its own personality, even though Tobey provides no distinguishing details. He succeeds in giving this locale both a cosmic and a human dimension. This points to an inherent paradox in small western towns: they are provincial, even backwards when judged against the largest cities, but at the same time they embody vitality and a potential for growth.

GUY ANDERSON ■

Guy Anderson (born 1906) was another founding member of the Northwest School of painters. A native to Washington, Anderson was a close friend of Morris Graves since 1929 and maintained friendships with Callahan and Tobey since the 1930s. In 1955 he settled in the little Washington fishing village of La Conner, an area that he had visited regularly since 1935. Bruce Guenther described this place as a "low, long, horizontal landscape" that was "surrounded by the wide, fertile fields of the Skagit Valley and crisscrossed by irrigation canals and a tidewater channel flowing into northern Puget Sound."[94] His works from the late 1950s were strongly influenced by this landscape. In *Deception Pass through Indian Country* of 1959, he employed a bold rhythmic patterning that permeates the entire scene. Abrupt zig-zag lines represent the force of the surging sea. Their staccato pulsation is picked up in the angular rocks on shore. This expressionist painting of seething energy captures the immediacy of nature one experiences along the rocky Northwest Coast.

Mark Tobey
Western Town. 1944
tempera
12 x 18¾ inches
Collection of Portland Art Museum,
bequest of Edith L. Feldenheimer
Photography by Guy Orcutt

Guy Anderson
Deception Pass through Indian Country.
1959
oil on paper on plywood
11 x 30⅜ inches
Collection of Seattle Art Museum, gift
of the Sidney and Anne Gerber
Collection
Photography by Chris Eden

LATE ABSTRACT EXPRESSIONISM AND THE WEST

RICHARD DIEBENKORN ■ Although in their formative years many of the Abstract Expressionists turned to the West for inspiration, they produced an art based on the unconscious that had little to do with geographic location. As a mature style Abstract Expressionism was tough, gritty and urban in outlook. Some artists, however, managed to utilize the style to reflect the unique qualities of the West.

For Richard Diebenkorn (born 1922) the West played a vital role in the development of his art. Younger than the first generation Abstract Expressionists, he created a personal variant of their gestural, Action Painting by studying examples of their mature work during the late 1940s. Living in New Mexico in the early 1950s, he began to translate the desert into free, improvisational compositions of muted color and free-formed line, as seen in his two paintings titled *Albuquerque*, both of 1951.

Diebenkorn was born in Palo Alto, California and studied at Stanford University. After serving in World War II, he enrolled at the California School of Fine Arts in San Francisco in 1946 and taught there from 1947 to 1950 where fellow faculty included Clyfford Still and Mark Rothko. He chose to pursue graduate studies at the University of New Mexico, Albuquerque, a decision based primarily upon photographs he had seen of the Southwest and on his general impression that he liked "the look of the place."[95]

Raymond Jonson was on the faculty and the two had rewarding conversations, but Diebenkorn was more interested in the spontaneous quality of gestural Abstract Expressionism than in Jonson's hard-edged geometric abstractions. Promoting the values he learned while in San Francisco, Diebenkorn broke with convention, warning against "prettiness" and facile imagery while advocating "ugliness," "sparseness" and "annihilating the image."

The abstractions from his Albuquerque period are as simple and uncompromising as the desert itself. Relying primarily on earth tones, he used color to mirror the general feel of the arid land. Gerald Nordland wrote that these works "had a preponderance of new elements: light, sand and flesh colors with a looping drawing, a rhythmic crudeness in the line, and deposits of color-form which developed out of revisions and transitions. There is little of conventional 'fine-handling' or seductive surfaces in these works."[96] Diebenkorn stated that "in Albuquerque I relaxed and began to think of natural forms in relation to my own feelings."[97]

His relaxation is seen in the quixotic, rambling line that moves unexpectedly as it traverses the canvas. Freed from the need to describe objects, this line became "quirkily elusive" as curator Maurice Tuchman wrote.[98] As seen in *Albuquerque*, in the collection of the Oklahoma Art Center, Diebenkorn's spontaneous drawing hints at objects formed through such slow natural processes as wind erosion.

Many of these paintings resemble the land seen from aerial views. *Albuquerque* of 1951, in a private collection, looks like a photograph of the desert taken from an airplane. Maurice Tuchman wrote of this work that "the white eruption of light at the right evokes some mysterious desert phenomenon. The lightninglike streak here seems related to Indian designs on pottery. This image is a fully-integrated landscape and still life — a table top shape is coincident with aerial landscape space."[99] Diebenkorn wanted an image that was frontal, direct and immediate. He had taken a flight over the desert in the spring of 1951 and was deeply moved by the experience: "The aerial view showed me a rich variety of ways of treating a flat plane — like flattened mud or paint. Forms operating in shallow depth reveal a huge range of possibilities available to the painter."[100] The impression that one is looking at an upright plane is strengthened by the lack of perspective or any hint of recession. He had eliminated blue at this time because it reminded him "too much of the spatial qualities in conventional landscapes."[101]

The New Mexico paintings of Richard Diebenkorn culminate the first phase in the relationship of Modern Art and the West. Earlier in the century, artists struggled to make sense of the region and then give coherent form to their perceptions. Diebenkorn effectively internalized this process. He embodied his understanding of the land directly in spontaneous gestures. Artists in the decades after World War II returned to the stereotypical images of the West as they reevaluated our cultural clichés in terms of a historic rise in mass communications.

Richard Diebenkorn
Albuquerque. 1951
oil on canvas
40½ x 50¼ inches
Collection of Oklahoma City Art
Museum
Photography by Sanford Mauldin

Richard Diebenkorn
Albuquerque. 1951
oil on canvas
38½ x 56¼ inches
Private collection
Photography by Janice Felgar

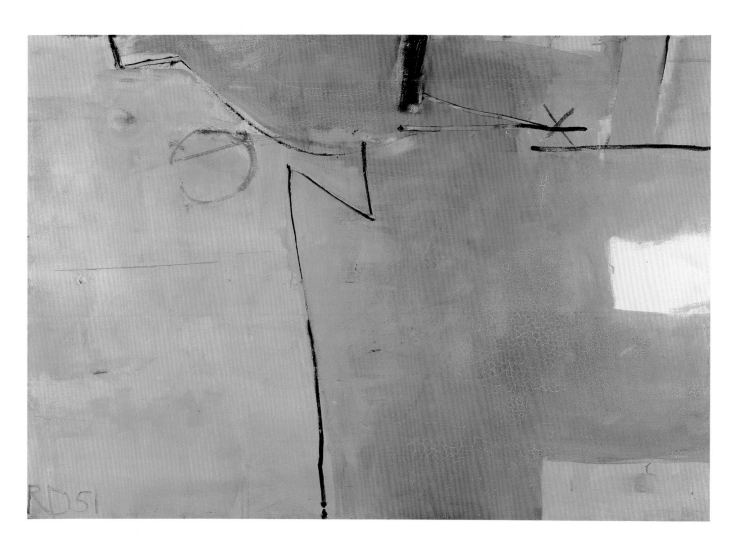

[1]Joseph S. Czestochowski, *Arthur B. Davies: A Catalogue Raisonné of the Prints* (Newark: University of Delaware Press, 1987), pp. 228-231.

[2]Martin S. Ackerman and Diane L. Ackerman, eds. *Arthur B. Davies: Essays on His Art, with Illustrations* (New York: Arco Publishing Company, 1974), n.p.

[3]Robert Hobbs, *Elliott Daingerfield: Retrospective Exhibition* (Charlotte, North Carolina: The Mint Museum of Art, 1971), p. 48.

[4]*Ibid.*

[5]*Ibid.*, p. 38.

[6]*Ibid.*, p. 40.

[7]Frederick C. Moffatt, *Arthur Wesley Dow (1857-1922)* (Washington D.C.: The National Collection of American Art, 1977), p. 117.

[8]Moffatt, p. 121.

[9]Wesley Burnside, *Maynard Dixon, Artist of the West* (Provo: Brigham Young University Press, 1974), p. 55.

[10]The California Academy of Sciences, *Maynard Dixon: Images of the Native American* (San Francisco: The California Academy of Sciences, 1981), p. 48.

[11]*Ibid.*

[12]Burnside, p. 74.

[13]California Academy of Sciences, p. 45.

[14]Patricia Janis Broder, *The American West: The Modern Vision* (Boston: New York Graphic Society, 1984), p. 136.

[15]Gail Scott, *Marsden Hartley* (New York: Abbeville Press, 1988), p. 51.

[16]Mahonri Sharp Young, *The Eight* (New York: Watson-Guptill Publications, 1973), p. 76.

[17]Broder, pp. 110-11.

[18]Ed Garman, *The Art of Raymond Jonson, Painter* (Albuquerque: University of New Mexico Press, 1976), p. 37.

[19]Georgia O'Keeffe, *Georgia O'Keeffe* (New York: The Viking Press, 1976), n.p.

[20]Katherine Kuh, *The Artist's Voice: Talks with Seventeen Artists* (New York: Harper & Row, 1962), p. 189.

[21]Jack Cowart and Juan Hamilton, *Georgia O'Keeffe: Art and Letters* (Washington, D.C.: The National Gallery of Art, 1987), p. 157.

[22]Calvin Tomkins, "The Rose in the Eye Looked Pretty Fine," *The New Yorker* (March 4, 1974), pp. 40-66.

[23]Anita Pollitzer, *A Woman on Paper: Georgia O'Keeffe* (New York: Simon & Schuster, 1988), pp. 148-49.

[24]Cowart and Hamilton, p. 155.

[25]Kuh, pp. 189-90.

[26]Cowart and Hamilton, p. 157.

[27]James Johnson Sweeney, *Stuart Davis* (New York: Museum of Modern Art, 1945), p. 15.

[28]Diane Kelder, ed., *Stuart Davis* (New York: Praeger Publishers, 1971), p. 41.

[29]*Ibid.*, p. 57.

[30]Lloyd Goodrich, *John Sloan* (New York: MacMillan, 1952), p. 52.

[31]Barbara Haskell, *Marsden Hartley* (New York: New York University Press, 1980), p. 58.

[32]Marsden Hartley, "Aesthetic Sincerity," *El Palacio* (December 9, 1918), p. 332.

[33]*Ibid.*

[34]Haskell, p. 58.

[35]*Ibid.*, p. 142, note 143.

[36]Hartley, p. 332.

[37]Garman, p. 68.

[38]*Ibid.*, p. 63.

[39]Sharyn Rohlfsen Udall, *Modernist Painting in New Mexico, 1913-1935* (Albuquerque: University of New Mexico Press, 1984), p. 97.

[40]Garman, p. 58-59.

[41]Dorothy Norman, *The Selected Writings of John Marin* (New York: Pellegrini & Cudahy, 1949), p. 128.

[42]*Ibid.*, p. 129.

[43]*Ibid.*, p. 128.

[44]*Ibid.*, p. 132.

[45]*Ibid.*, p. 135.

[46]*Ibid.*, p. 134.

[47]Cowart and Hamilton, p. 189.

[48]Georgia O'Keeffe, n.p.

[49]Sidney Lawrence, *Roger Brown* (New York: George Braziller, 1987), p. 96.

[50]Kuh, p. 194.

[51]Cowart and Hamilton, p. 203.

[52]Hirschl & Adler Galleries, *Georgia O'Keeffe: Selected Paintings and Works on Paper* (New York: Hirschl & Adler Galleries, 1986), n.p.

[53]*Georgia O'Keeffe: Exhibition of Oils and Pastels* (New York: An American Place, 1939).

[54]Kuh, p. 194.

[55]Cowart and Hamilton, p. 145.

[56]*Ibid.*, p. 155.

[57]*Ibid.*, p. 201.

[58]Georgia O'Keeffe, n.p.

[59]"Reviews and Previews: George L.K. Morris," *Artnews* (January 1955).

[60]Douglas Dreishpoon, *New York Cubists: Works by A.E. Gallatin, George L.K. Morris, and Charles G. Shaw from the Thirties and Forties* (New York: Hirschl & Adler Galleries, 1988), pp. 23, 43.

[61]George L.K. Morris, "On the Mechanics of Abstract Painting," *Partisan Review* 8 (September-October 1941), p. 410.

[62]*Ibid.*

[63]Garman, p. 87.

[64]Phil Dike, "Phil Dike: He Captures the Scale of the West," *American Artist* (November 1940), p. 20.

[65]*Ibid.*, p. 19.

[66]*Ibid.*, p. 20.

[67]*Ibid.*, p. 19.

[68]*Ibid.*, pp. 19-20.

[69]*Ibid.*, p. 20.

[70]Lew Davis, "Commentary," in Jon H. Hopkins, *The Art of Lew Davis: A 40 Year Retrospective* (Flagstaff, AZ: Northland Press, 1970), pp. 10-11.981), p. 21.

[71]Sanford Hirsch and Mary Davis MacNaughton, *Adolph Gottlieb: A Retrospective* (New York: The Art Publisher, 1981), p. 21.

[72]Adolph Gottlieb to Paul Bodin, December 2, 1937, the Adolph Gottlieb Foundation.

[73]*Ibid.*, December 29, 1937.

[74]*Ibid.*, January 23, 1938.

[75]*Ibid.*, March 22, 1938.

[76]Adolph Gottlieb, "My Painting," *Arts and Architecture* (Spring 1951), p. 21.

[77]Barbara Rose, "Painters of a Flaming Vision: El Greco and Jackson Pollock," *Vogue* (December 1981), p. 334.

[78]Howard Putzel, "Jackson Pollock," *Arts and Architecture* (February 1944), p. 44.

[79]John O'Brian, ed. *Clement Greenberg: The Collected Essays and Criticism*, vol. I (Chicago: The University of Chicago Press, 1988), p. 165.

[80]Joanne Kuebler and Robert Hobbs, *Richard Pousette-Dart* (Indianapolis: The Indianapolis Museum of Art, 1990), p. 22.

[81]Ibid., p. 22.

[82]Ibid., p. 24.

[83]Judith Higgins, "Pousette-Dart's Windows into the Unknowing," *Artnews* (January 1987), pp. 111-12.

[84]Ibid., p. 26.

[85]Lucy R. Lippard, "Richard Pousette-Dart" *Artforum* (January 1975), p. 53.

[86]Ibid., pp. 51-53.

[87]Ann Gibson, "Painting Outside the Paradigm: Indian Space," *Arts* (February 1983), p. 100.

[88]Lee Nordness, ed. *Art USA Now* (New York: Viking Press, 1962), p. 208.

[89]Henry Geldzahler, *American Painting in the 20th Century* (New York: The Metropolitan Museum of Art, 1965), p. 165.

[90]Martha Kingsbury, *Northwest Traditions* (Seattle: Seattle Art Museum, 1978), pp. 44-45.

[91]Ray Kass, *Morris Graves: Vision of the Inner Eye* (New York: George Braziller, 1983), p. 47.

[92]Kingsbury, p. 50.

[93]Douglas Davis, *Newsweek* (July 29, 1974), p. 58.

[94]Bruce Guenther, *Guy Anderson* (Seattle: Francine Seders Gallery, 1986), p. 94.

[95]Robert T. Buck Jr., *Richard Diebenkorn: Paintings and Drawings, 1943-1976* (Buffalo: Albright-Knox Gallery, 1976), p. 13.

[96]Gerald Nordland, *Richard Diebenkorn* (Washington, D.C.: Washington Gallery of Modern Art, 1964), p. 10.

[97]Ibid., p. 11.

[98]Buck, p. 15.

[99]Ibid., p. 16.

[100]Gerald Nordland, *Richard Diebenkorn* (New York: Rizzoli, 1987), p. 43.

[101]Buck, p. 15.

Roy Lichtenstein
Cowboy on Bronco. 1953
oil on canvas
32 x 26 inches
Collection of The Butler Institute of
American Art, Youngstown, Ohio

A Contemporary Transformation

The history of the American West is constantly being written by scholars who present fresh insight and facts that incorporate diverse cultural viewpoints. Art also has played a significant role in portraying the West's history and often is cited as a record of the time. Presenting distortions based on individual perspectives, it further illuminates and interprets fact and fiction.

From Pop Art of the 1960s to Appropriated Art of the 1990s, contemporary artists have been fascinated by the western landscape and the changing perspectives of its motifs. Myriad images have triggered their concepts of the West, some real and some purely imaginary. Pop Art, with "superstars" Roy Lichtenstein and Andy Warhol, has been among the most visible of the twentieth-century art movements to interpret the West. The works, referred to as *low art*, have borrowed subject matter from twentieth-century routines and images. "Pop Art, in its original form, was a polemic against elite views of art in which uniqueness is a metaphor of the aristocratic and contemplation the only proper response to art."[1]

Transforming the Western Image explores how artists of the late twentieth century have created new and nurtured existing images of the West. Some of America's most enduring and treasured myths have persevered through their multiple investigations. This exhibition suggests the impact of their inquiries that have both documented and stretched facts from some of the West's most enigmatic memoirs.

ROY LICHTENSTEIN ■ Roy Lichtenstein (born 1923) was a leading figure in the post-World War II American hegemony that had abandoned utopian ideals well before Pop Art emerged in 1960. Lichtenstein was fascinated with Americana, and from 1951 to 1957 his art dealt with the American West and various Native American themes. During this period he lived in Cleveland and worked as a graphic and engineering draftsman, window designer and sheet-metal designer while painting satires of cowboys and Indians, such as *Cowboy on Bronco* of 1953 and *The Straight Shooter*, circa 1956. The cowboys are freely restyled from reproductions of bronco busters by Edward Borein and Frederic Remington that Roy Lichtenstein had seen.[2] He combined and restated elements of the horse, corral, chaps, spurs, boots, hat and plaid shirt. A Cubist version is *Two Sioux*, circa 1952 to 1954, revealing more abstract and increasingly sophisticated work.

Postwar consumerism and the return to recognizable imagery were natural reactions to Abstract Expressionism. Pop artists rejected spiritual and psychological elements in favor of a more humorous and sarcastic approach to art and life. Lichtenstein became well known around 1961 as the Pop artist who transformed comic strips and commercial advertisements into Benday-dot paintings.

Lichtenstein believes in composition as the balance of contrasting but compatible forms, in which size, direction, and color can be related; in which colors compensate for cool, in which curves ameliorate right angles, and in which details enliven large spaces. His work until the mid-1960s is constructed on these principles, and his accommodation of far-out image-source with academic picture-building is an engaging aspect of Pop art's play with ambiguous sign-systems. It is not, as formalist critics presumed, a sign of weakness but one of doubt operating equally corrosively in two directions.[3]

Lichtenstein is a satirist of the visual cliché who loves the banal and transforms it into *high art*. Instead of the organic forms, psychologically unconscious marks and psychic self-expression of

Roy Lichtenstein
The Straight Shooter. ca. 1956
oil on canvas
16 x 13½ inches
Collection of The Butler Institute of
American Art, Youngstown, Ohio

Roy Lichtenstein
Two Sioux. ca. 1952-1954
oil on canvas
30 x 22 inches
Private collection
Photography by Philipp Scholz
Rittermann

Roy Lichtenstein
American Indian Theme III. 1980
woodcut
35 x 27 inches
Collection of
Palm Springs Desert Museum,
purchase with funds provided by
the Walter N. Marks Graphics Fund
Photography by Larry Reynolds

74

Roy Lichtenstein
American Indian Theme VI. 1980
woodcut
37½ x 50¼ inches
Collection of Walker Art Center,
Minneapolis, Tyler Graphics Archive

Abstract Expressionism, he gives us distance instead of intimacy, cool instead of passion and the centrality of the common instead of the heroic. Lichtenstein cultivates the crude, mass-produced and anonymous image.

Since the 1960s he has been primarily interested in painting parodies of art-historical masters, series after series of paintings, drawings, prints and sculpture in Cubist, Purist, Art Deco, Futurist, Surrealist, Abstract Expressionist and German Expressionist styles. Lichtenstein boldly challenged high art values by contrasting them with a vernacular mode and by introducing references to art-historical styles and his own artistic development. Within a chosen theme, he remains constantly inventive, masterfully creating powerful compositions by manipulating color, line, form, scale and a vast repertoire of images. His is an art of reference: he borrows material from any artist to suit his needs.

In 1977 Lichtenstein returned to American culture themes creating paintings about the American Indian. "His 1977–79 Surrealist works evolved to a formal point close to his earlier Indian themes. . . . Feeling that Indian forms establish a historical base for American art, . . . the artist again willfully demonstrated in 1979 the dramatic flexibility of his style and choice by personally reinterpreting these subjects."[4] Lichtenstein's Indian designs are often based on "dream revelations," of ritual Indian art and "influences of Surrealist art, like the 1940s pictographs of Adolph Gottlieb and the Indian-Surrealist works of Max Ernst."[5] His complicated Surrealist-style images of Native American themes exemplify one of Lichtenstein's more durable and compelling artistic concerns.

Lichtenstein has taken the implied meanings of Surrealism within our European cultural heritage and magnified them through the ceremonial symbolism of the Indian. The explicit results produce paintings that are provocative on several levels, ranging from an appreciation of cultural qualities and the political minority issues of today's American Indian movement, to the debased Indian motifs in such stereotyped vehicles as Cowboy-and-Indian movies and the tourist trading-post totem pole which Lichtenstein had in his studio in the early 1960s.[6]

In 1980 he completed a woodcut series and a group of intaglio prints at Tyler Graphics on the theme of the American Indian. *American Indian Theme III* and *VI* are Surrealist distortions and interpretations of his own abstract Indian paintings of the 1950s. This evolution of images borrowed from his own pre-Pop Art, which had evolved from reproductions, are personal, surreal parodies of recent art history. In them Lichtenstein has combined the quintessential clichés of the American Indian. Denying his source material's original intentions, he combines tribal and geographic areas, and he skillfully refigures

images and overlapping forms found in kachinas, pottery, feathers, headdresses, beadwork and blankets. These works are nearly illegible in an imaginative narrative with a jarring mixture of scale.

ANDY WARHOL ■ Like Lichtenstein, Andy Warhol (1928–1987) derived many of his early paintings directly from comic strips, but then he sought inspiration from the commercial world. Warhol became the ultimate Pop artist. By taking objects and images from popular culture and reproducing them in silkscreen, Warhol documented the vernacular, the famous, the frightening and the tragic. His prophetically uncanny foresight resulted in the use of enormously popular images—Marilyn Monroe and Elvis Presley—and scenes of tension or anguish—the electric chair, race riots, automobile accidents—that he could manipulate and dematerialize into silkscreened fantasies. Emotion was defused through repetition.

When Andy Warhol, Roy Lichtenstein and Robert Rauschenberg first used mechanically produced commercial imagery in their paintings, they employed technological procedures including the dye transfer process and silkscreen printing. Gestural identity and original expression were challenged. Using mechanical means of reproduction, Pop Art became a new art form and immediately became the ultimate object for collection. "They (Pop artists) took up mechanical methods and mechanistic images to intensify not to discredit the latter-day aura of mass-produced images and objects, these presences whose encircling, relentless glamour has an authenticity that they above all never doubted."[7]

The silkscreen process, a technique in which a stencil is created by either a hand-drawn, cut or photographic process, was appropriate for Warhol's serial images of portraits and commercial products. It produces flat and sharply defined outlines as in *Portrait of Dennis Hopper* of 1970. Fascinated by the cool, impersonal reproductions of famous emotive characters, Andy Warhol silkscreened photographically enlarged reproductions onto canvases. In this way he could capture the evanescent personality of a cowboy in the portrait of Dennis Hopper.

Lichtenstein's and Warhol's cowboy and Native American paintings are both imaginary and real. They refer to established motifs of both objects and images—usually reproductions—that revert to become images in their own right. The silkscreen process enabled Warhol to investigate and represent any media language that was interdependent with the camera. Warhol photographed North American Indian artifacts, kachinas, shields and ceremonial masks at the Museum of the American Indian in New York, and he recreated them along with portraits in his 1986 *Cowboys and Indians* portfolio. Recalling turn-of-the-century postcards, his objects and portraits play on the myths of the American West. Warhol's Hollywood version of the period is represented by strong and aggressive cowboys: *John Wayne, Teddy Roosevelt,*

Andy Warhol
Portrait of Dennis Hopper. 1970
acrylic on canvas
40¼ x 39⅞ inches
Collection of an anonymous lady

Andy Warhol
Mother and Child (#383). 1986
serigraph
36 x 36 inches
Courtesy of Kent M. Klineman

General Custer and *Annie Oakley*. Images of vulnerable Indians include a squaw and papoose in *Mother and Child* and an aged *Geronimo*.

By focusing attention on the fragile heritage of the Indian nation, Warhol is not simply another celebrity promoting a cause. He has long had an interest in Native American art, not only because it shares with his own a graphic, decorative treatment of color and line, but because Indian art is as outer-directed and reactive to the environment as the artistic products of Warhol's notoriously cool gaze. As the most radical proponent of Pop, that diamond-hard, quintessentially American contribution to contemporary art and world culture, it is altogether appropriate that Warhol should be drawn to the other indigenous American culture. Both Warhol and the Indians had an independent origin in America, and both have had to resist the European influence to maintain the strength and purity of their tribal cultures. In the *Cowboys and Indians*, Andy discovers his roots.[8]

Warhol's preferred style used images in an inconsistent yet characteristic way. He was one of many who rendered subject matter with duality and ambiguity. Red Grooms plays a comparable game with his art.

RED GROOMS ■ To maintain a sense of unreality in a dependable relationship with reality, Lichtenstein, Warhol and Red Grooms (born 1937) have provided a style for many tastes. Grooms's cunning style not only entertains but presents an excitable balance between direct documentation and bitter ironic wit. His cartoon-like renditions, exaggerated perspectives and willful distortions rouse an immediate humorous reaction. For the message to be conveyed, however, a second look is often required, which Grooms calls his "double nature."

He uses highly charged imagery mingled with clichés of popular culture and self-righteous values of the art world. A comic approach plays off mundane activities and harsh realities. With keen attention to visual facts and political savvy, Grooms cannily documents the human condition. "Grooms's art is so deeply immersed in the traffic that surges through the streets, across the television screen and through our overburdened synapses that it is never a surprise to see current events helping him complete the meaning of his images."[9]

Grooms's artistic expression includes painting, sculpture, print-making, large-scale environmental pieces and movies. In the past thirty years critics have sought to clarify the relationship of Red Grooms to the art of his times, his work not seeming to fit within a convenient historical niche. In fact, Grooms consciously works against the art world's orderly groupings by being too much a humanist for

Red Grooms
Great Western Act. 1971
oil on masonite
23½ x 35 inches
Collection of Red Grooms, New York

Red Grooms
Shoot-out. 1980
painted bronze
11½ x 22 x 9¼ inches
Private collection

Pop and too fundamentally fun-loving to be a scathing satirist.

Grooms developed his active expressionist style in the late 1950s and early 1960s when he presented performance pieces or Happenings, using props and "stick-outs" that featured an incipient spirit named Ruckus. Grooms joined artists Allan Kaprow, Claes Oldenburg, Jim Dine and others who often involve the audience in their creation, breaking down subject matter barriers and their own isolation. After creating and performing *The Burning Building*, a Happening in 1959, Grooms so enjoyed the audience that he easily gravitated to filmmaking. Through his 1961 animated movies, he communicates rich, elaborate and zany ideas that he was unable to express in other media. Collaborating with fellow artists, friends and family, Grooms is able to produce work that broadens and exploits the narrative dimensions of his art, while representing a cross-section of history.

After participating in Happenings and making a number of imaginative and personal films, Grooms found that he enjoyed the collaboration of an audience to such a degree that he decided to focus on interactive constructions of walk-through, three-dimensional environments. Today he is recognized as a pioneer of site-specific sculpture and installation art. In *Ruckus Rodeo* of 1976 the West was the theme for one of the walk-through, room-sized environments he has built with his wife and the Ruckus Construction Company (a consortium of local artists, students and other enthusiasts). Other environments are *City of Chicago* of 1967, *Discount Store* of 1970, *Ruckus Manhattan* of 1975, and *A Philadelphia Cornucopia* of 1982.

A copious amount of energy exists in Grooms's works with the sub-

jects selected and the predictably frantic fashion in which he treats them. The overactive environmental play of racial and social stereotypes in the West and urban centers become points of departure for his personal refashioning of American history. The sum of these parts is "Groomstown," a cartoon of a city bounded on one side by a rodeo and with an art museum at its nucleus.[10]

Part historian and part painter of contemporary life, the ambitious and prolific artist includes recurring self-portraits in his visionary works, along with group portraits of family and friends and series of paintings on the old West. In *Great Western Act* of 1971, Grooms's whimsical insights and attitudes come into distorted play in caricatured rodeo performers and their audience. Reminiscent of an old Wild West Show, a cowboy, cowgirl and bucking bronco are central characters in this painted "theater." A stage crowded with wacky characters in comic action-drama convey aspiring youthfulness and aggression with rough awkwardness.

One of the attributes of Grooms's working style is his ability to revive his subject matter with each new medium used; materials selected prompt distinct stylistic handling. As part of his visual commentary on the American "character," Grooms has created painted bronzes depicting the West, football games and other generically American subjects. *Shoot-Out* of 1980, painted in his strident reds, purples, Day-Glo pinks and blues, echoes the immediacy of folk art. The fighting cowboy and Indian, in a wagon surrounded by suspended bullets and arrows, are presented in Grooms's typically distorted and humorous manner.

LLYN FOULKES ■ Art critic Peter Plagens has described Llyn Foulkes (born 1934) as one who "worked in the cracks between assemblage and hard-core Pop."[11] During Foulkes's thirty-year career, a personal and fiercely individual vision has developed from the Dada/Surrealist tradition.

Along with artists who personified the developing Los Angeles art sensibility in the 1960s, Foulkes had ideas common to Pop Art. By combining humorous and ironic elements with a California Pop attitude and a disregard for high art seriousness, his ideas bent stylistic rules. Foulkes has consistently produced art that conflicts with the cherished values of the established art world; using both physical and psychological bravado, he demolishes style, taste and fashion. He says—

> I refuse and will always refuse to become a product for the sake of ART. I am still challenging my securities to find if they are real and my mistakes to find if they are mistakes. I believe in the most direct approach to art. If it is going to live, it must take from everything (including itself). It must find a valid reason for existing in an overstimulated world.[12]

While attending the Chouinard Art Institute in Los Angeles from 1957 to 1959, Foulkes adopted the then-dominant influence of Abstract Expressionism, but he placed recognizable imagery in work. Embedding figurative imagery in dark, gestural brushwork, he incorporated structural elements that intensified the work's symbolic meaning. Although drawn to New York Neo-Dadaism, his experiments with collage and construction personified California assemblage and Pop Art. In early assemblages made between 1959 and 1962, he scrambled images of political caricature and adhered found objects to their surfaces.

Foulkes discovered early on that the photo image satisfied his need to work with both reality and abstraction; it shared concerns of surface, texture, object, composition, value and two-dimensionality. Foulkes did not reproduce a photograph exactly; his paintings-cum-photographs resulted in the "postcard" images produced between 1963 and 1970 in which he used western landscapes with mountain ranges, hills and rock formations inspired by Los Angeles suburbs— Pasadena, Highland Park and Eagle Rock. The landscapes simulated photography through manipulation of the flat surface of the picture plane and illusionistic spaces. Using a technique of controlled scrubbing of the pigment, the simulated photographic rock formations or landscapes were combined with either flat or textured areas. The photographic approach was a common factor from painting to painting, however it was not the essence of Foulkes's work as it was with Warhol's.

Using landscape as a point of departure, stony hills and buttes

suggest silhouettes of heads, torsos and anthropomorphic forms. These paintings incorporate a vocabulary of images: horizontal and vertical stripes, stencils of numbers or words (including the word "Post Card") and handwritten inscriptions or titles of dedication. Although these do not associate with Pop media imagery as much as they probe and ponder psychological depths, the perception of viewing the blurred picture (as if from a speeding vehicle) relates to a Pop sensibility. The relationship is enhanced by the mimicry of photographic source and process; Foulkes paints each mountain individually with slight, "handmade" differences inherent to such handling. In this body of work Foulkes created a synthesis of Pop/Minimal coolness with a disconcerting Surrealist intensity.

> The mountain views in *The Page* (1963) look as if they were documents from some geological research project exhumed from a forgotten archive. The simulation of photography functions less as a pop reference to flat reproduction techniques and more as a re-creation of anonymous autobiography. Moreover, Foulkes scribbles on the surface of the canvas as if the "photographer" had scratched, dated, and annotated his work, giving the bleak landscape a certain intimacy. The "hand" remains anonymous as if it flowed from the "people" as a whole. The sense of a collective voice is echoed in the single readable phrase: "This picture and These volumes are dedicated to the American." The source of this sentence-fragment is Ulysses S. Grant's frontispiece dedication from his memoirs, which begins, "These volumes are dedicated to the American soldier and sailor." Foulkes's use of this Civil War reference is grounded in his fascination with American history.[13]

Foulkes sets up other pictorial tricks as well, such as the unusual manner in which he frames his images. As an ironic device he often uses only one vertical and horizontal framing element, omitting the others. The frame as a primary element is further developed in a series of portraits, begun in 1973. In small, mixed-media portraits of psychological horror, typical subjects are authoritative male figures from military, corporate, bureaucratic or political arenas. Framed with construction materials, they symbolize the social structures selected to protect them. "Foulkes' portraits are so powerful that they bear almost no relationship to the bogus emotion and smart anxiety of much recent expressionistic art. Instead, they beat a path to the abyss of German Expressionism and to Goya's black paintings."[14]

In works of the 1980s, Foulkes constructs elaborate portraits and diagrammatic pieces by building each image through accumulation. In *The Last Outpost* of 1983, a witty and imaginative autobiographical work, movie and television characters are combined to recall his childhood dream of becoming a cowboy. By combining the Lone Ranger

Llyn Foulkes
The Page. 1963
oil on canvas
87½ x 84 inches
Collection of The Oakland Museum,
gift of Anonymous Donor

Llyn Foulkes
The Last Outpost. 1983
mixed media assemblage
81 x 108 x 5 inches
Collection of
Palm Springs Desert Museum,
purchase with funds provided by
the Contemporary Art Council, 1989
Photography by Marc Glassman

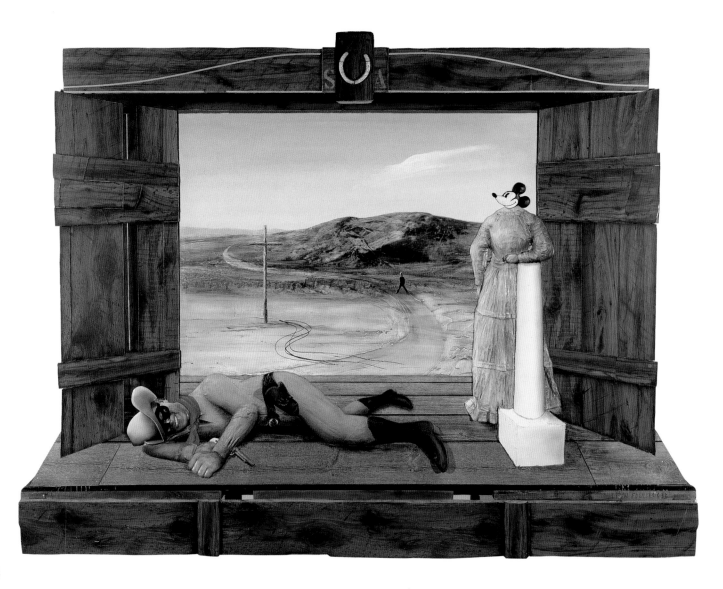

and Minnie Mouse, the artist invites us to reflect on what is real and imaginary. Foulkes wrote a lyric to accompany this work:

My father told me if I ate my vegetables and cleaned my plate that I could be a cowboy just like the Lone Ranger. My mother told me if I took my medicine and read my book that I could be a cowboy just like the Lone Ranger. I got a rifle, I got a pony, my mother said I could play outside if I finished my macaroni. I shot the postman in the head and rode away cause he was dead then I sang a song just like the Lone Ranger.

I've been a cowboy ever since the age of ten and I've been out mendin' fences since I can't remember when. I never thought that a cowboy was a guy who rode the range. I guess I'll have to go out further west to change. Southern California is a place I ought to go, where the cowboys yip and the coyotes howl and they all make a lot of dough. Southern California won't you wait for me. It's the only place where a cowboy can really be free.

Now that I have been upon the silver screen it doesn't really mean that I'm a cowboy. I wish that I were only ten it was so simple then to be a cowboy. You could ride the western range it never seemed to change it never was untrue. You could hide beneath the stars and shoot at passing cars like real cowboys do.[15]

In the analysis of this work by *Los Angeles Times* critic William Wilson, he states:

In the foreground lies the Lone Ranger, mortally wounded by a shot fired by a small boy in the distance. Nearby stands a pouter-pigeon Victorian woman with a Mickey [sic] Mouse head. The masked man grins happily, giving the grim scene an almost lyric make-believe tone. It is the child's version of Freud's "Murder the Father" scenario. What the boy does not realize is that the imaginative assassination of the old man means that he has to grow up and abandon his fantasies.[16]

Foulkes uses the landscape and figure in an "altered state" to explore the emotional tension between actual and illusionistic space. This investigation recalls his postcard paintings and Pop orientation of the 1960s.

William Allan
Shadow Repair for the Western Man.
1970
oil on canvas
90 x 114 inches
Collection of University Art Museum,
University of California at Berkeley
Photography by Benjamin Blackwell

WILLIAM ALLAN ■ Like Foulkes, William Allan (born 1936) is an artist who does not conform to the *de rigueur* of the art world. Referring to himself as a poet rather than an artist, Allan approaches art with his feelings about life and "things that one understands as a poet, but which aren't logical." The social rigors of the gallery scene, including the public viewing of his work, have little meaning for Allan. Rather than dealing with the perplexing and formal problem-solving activities as do many artists, he is interested in the enigmatic nature of human existence, willfully foregoing a relationship with collectors or critics.

For twenty-five years, Allan's work has been consistently autobiographical; subjects are drawn from real objects and selected for their association with particular events. Images are inspired by dreams, experiences, stories, friends and family.

Allan's box constructions and assemblages of the 1960s reveal arcane stories with visual puns and word play, an interpretation of ideas that ridicule the high art pretenses of the 1950s. Along with a number of artists who were students or instructors at the San Francisco Art Institute and the University of California, Davis, Allan explored ironic twists and created art guided by anthropomorphic and passionate characteristics (later termed Funk Art after Peter Selz's definition of the term in his 1967 Berkeley exhibition). They challenged reality, emphasized process, distrusted too much logic and thought and contradicted commonly held beliefs. They produced bodies of work to articulate the range of human experience, from the poetic, transcendent and unexpected, to the commonplace. Aggressively mocking in tone and often materially crude, Funk Art became California's most distinctive contribution to America's Pop Art movement.

More specific to the West, the vast and illusionistic paintings of clouds and landscapes of the late 1960s and 1970s are Allan's complex story paintings with enigmatic titles. Proclaimed seriousness is confirmed by immense scale, magnificent paint handling and Surrealistic juxtaposition. In *Shadow Repair for the Western Man* of 1970, the western American man's energy is described in motionless detail — an empty pair of Levi's with old shoes on the seat of a broken chair overlooking a majestic snowcapped mountain range in a breathless air.

The notion of repairing things, which fascinated Allan to the point of organizing the San Francisco "Repair Show" in 1969 (to which an enormous number of artists contributed works), is a theme which is at once physical, psychological, structural, metaphorical. It is also, in this painting, a personal allegory for the artist. Partly, it is about a friend who revealed to Allan "my other, more delicate side, something that allowed me to accept my provincialism, being a Western man." The painting, Allan says, also reflects "the images we all have of ourselves that are

being held up by shaky props. We don't want to love all that, but it's preposterous to tear it down. Even if it's shaky, it should be a monument." There is a bizarre, rather surreal juxtaposition of human artifacts and natural phenomena in the painting. The human element is vulnerable, while the mountains are cold and remote. "Nature is where you go to get the shit scared out of you," he says. "The mountains are bigger than any comprehension we can have about them. I made them cold, to be about respect."[17]

An avid fisherman, Allan's massive landscapes are followed by intimate watercolors of salmon and trout. Explanations of life experiences touched by fish and the fisherman, he records natural phenomena with exquisite detail. Accompanied by simple stories, they become objects for contemplation. "Allan describes his paintings in much the same metaphysical way he describes fishing. Although he acknowledges a vast difference between the two activities, he does not consider one more important than the other."[18]

Recent paintings contain the subdued Surrealism, reference to personal events and majestic detail of his earlier works and remain artistically illuminating. His art is a vehicle to sort through his own experience in relation to the mysterious and unaccountable events, ideas and feelings of the human condition.

ROBERT ARNESON ■ Since the Funk Art movement of the 1960s, Robert Arneson (born 1930) has been active in California's ceramic sculpture movement, becoming known as one of the world's leading humorists via the clay medium. Instrumental in taking clay from its traditional utilitarian role and bringing it into the realm of fine art sculpture, he has bridged the gap between caricature and serious portraiture, questioning with sardonic humor the meaning of art. Witty drawings and sculptures are saturated with visual and verbal puns. "In terms of art movements he has critiqued High Renaissance art, Dada, Surrealism, Abstract Expressionism, Pop Art, Earth Art, Body Art, Minimalism, Color Field Paintings, and Neo-Expressionism. He has made fun of the excesses of modern art at the same time he has upheld its ability to deal with feeling and also symbolize the political struggles of the present-day world."[19]

Through the late 1960s and 1970s, Arneson expanded his style well beyond his earlier ceramics. Combining personal and humorous comments with art historical references, his work remains distinct from mainstream formalist styles. During this time he also concentrated on autobiographical works, the vast majority of which have been self-portraits in clay. In various guises and attitudes, he pushes self-portraiture to an extreme. He says he distorts his own features to develop a total characterization of emotion and reaction that captures a

Robert Arneson
Last of the Great White Buffalo Hunters.
1987
wood, fixall, ceramic, paint, leather
57 x 74 x 32 inches
Collection of Denver Art Museum,
funds from Colorado Contemporary
Collectors and National Endowment for
the Arts Museum Purchase Program

Jackson Pollock
The She-Wolf. 1943
oil, gouache and plaster on canvas
41⅞ x 67 inches
Collection of The Museum of Modern
Art, New York, purchase

frozen moment in time. These works have multiple meanings; titles are given as clues to narrative references.

In a similar exploration, he has created portraits of other artists—Peter Voulkos, Vincent van Gogh, Marcel Duchamp, Francis Bacon, Pablo Picasso, Jackson Pollock and others, interpreting their likeness to capture and relate events or anecdotes. On each bust Arneson comments about twentieth-century art, life and humanity with witty graffiti. Satire is integral, and he achieves expressive statements through biting distortions.

> Arneson's work is a fascinating record of the period from the 1960s to the present because it accepts the limits of the artist's ego and recognizes all art as part of an ongoing dialogue with other works of art and also with the world at large. It relies on bad taste and humor to keep this critical stance from becoming merely academic, and in the process it continues to puzzle, shock, and delight viewers who find its approach truthful and lively.[20]

In the 1980s his works became more serious, often infused with threatening undercurrents of nuclear holocaust. Images of human heads, all victims of atomic warfare, reveal Arneson's brand of black humor. Perpetrators, officers and leaders in uniform, all portrayed as destructive, are depicted as venomous and gross. Expressing disdain for militant authority figures, Arneson makes powerful comic statements to express concerns for a peaceful world.

About this same time, Arneson began a series of portraits of Jackson Pollock, for Arneson the archetypal artist-victim-survivor. *Last of the Great White Buffalo Hunters* of 1987 is a tribute to Wyoming-born Pollock and his 1943 painting entitled *The She-Wolf*. On the flanks of Arneson's emaciated wolf is a sensitive relief portrait of Pollock, who died at forty-four in a car accident. In this ceramic sculpture, Arneson has placed images of the Old West and symbols from *The She-Wolf* that represent the tragedy of senseless extinction and loss of life. Pollock wears a buffalo headdress with a message-scrawled piece of leather hanging from one of the horns. Remains of animals and a human skull, upon which Arneson signed his name, are placed at the foot of the wolf among piles of rock and earth.

> Arneson's identification with Pollock recovers the sense of Pollock as victim, as disturbed, even deeply pathological—which is the only way to be "authentic" in the modern world. For Arneson, Pollock represents the socially rejected, isolated individual in American life. His Pollock is the brutalized victim-survivor—which is what Arneson suggests the authentic

artist is—of the modern, nuclear age. For Arneson, Pollock's life, like that of van Gogh, was in itself "political"—both artists were, in Artaud's phrase, suicided by society. And it is this suffering (if sometimes tough-looking) Pollock that Arneson invokes, with vigorous painterly gestures which seem to constitute Pollock's figure rather than be superimposed on it. This gesturalism fully communicates Pollock's anguish and self-dramatization.[21]

H.C. WESTERMANN ■ A well-known creator of poetic sculpture, H.C. Westermann (1922-1982) has shared an idiosyncratic view of the world through watercolors, letters and drawings sent as letters. Although critics have compared his art to Dada, Surrealism and the Funk Art movement, Westermann's fierce individualism prohibits easy classification. He uses the visual pun, found objects, humor and accidental effects of Surrealism with a Pop-like use of caricature and cartoons. Because both made enigmatic boxes, Westermann's work has often been compared to Joseph Cornell's. Westermann's outlook was distinctly an American one, however, reflecting the childhood experiences of growing up in California in the 1930s.

> His art necessarily reflects the artistic climate in which it developed, but he remains essentially a loner with a unique sensibility. He has set an example for later generations as a true individual whose art has never been motivated by formal or theoretical ideologies but instead manifested his own philosophy and life experiences. By pursuing a private sense of achievement based on a personal code of ethics and behavior, his objects violate conventional notions of art historical theory and taste.[22]

Rife with illusions and references to highly personal subject matter, his ideas and images are a result of an interaction between memory, external reality, experience and imagination. Self-caricatures depict a variety of guises to express particular events, but meanings remain elusive as Westermann intended.

> What makes Westermann unusual as an artist is his ability to so graphically visualize the relationship of the self to the other. The dream-like quality of his imagery and his fertile mixture of word and image suggest the movement of the subconscious outward as it seeks its features in forms that resist limitations imposed by ordinary social and cultural boundaries and so resist easy interpretation. Westermann's art is a celebration of the self's ability to prevail even as it incorporates the outsider into itself.[23]

H.C. Westermann
The Stranger. 1978
watercolor on paper
22¹⁄₁₆ x 30¾ inches
Collection of Seattle Art Museum, gift
of the Contemporary Art Council
Photography by Susan Dirk

Westermann's assured placement of seemingly incongruous parts heightens their mysterious references. While the pieces appear at first to be merely wacky and whimsical art, at closer examination his pieces are more complex and of vastly greater aesthetic and philosophical significance. Subtle statements rely on the viewer's alertness and ingenuity for comprehension.

Cliff of 1970, a meticulously crafted glass-enclosed diorama, revolves around man's helplessness in the face of events and forces larger than himself. A cowboy astride his horse is pushed to a cliff by attacking Indians, portrayed in a painting by Westermann's wife, Joanna Beall. Westermann placed a mirror, symbolic of metaphysical reflection, at the base of the cliff, counter-balancing vulnerability with imminent catastrophe. The horse on the cliff is a play of words drawn from the artist's name, Horace Cliff.

Westermann's self-projection extends to animals, often feral-looking dogs. In *The Stranger*, a watercolor of 1978, the animal form appears with Westermann in the shadow of a human engaged in struggle. The dog is ostracized to the margins of civilization so that society is not threatened by its actions. The artist renders the melodrama of adventure and experience through a mixture of scale, perspective, verbal and visual puns and the use of flatly applied color.

WAYNE THIEBAUD ■ The work of Wayne Thiebaud (born 1920), has been linked to a number of art movements, including Pop Art, New Realism and American Realism. Not easily characterized to clearly defined art movements or issues, Thiebaud arrived on the New York art scene in the 1960s during the emergence of Pop Art. Although he did not participate in the movement, he was quickly identified with it. He painted subject matter with mass recognition — foodstuffs, pinball machines and dime store lipsticks, choices made as a means to avoid belabored art approaches.

Notable distinctions exist, however, between Pop artists and Thiebaud. While his Pop colleagues painted with seemingly detached sensibilities, Thiebaud's canvases convey meticulous handling of materials and purposeful image making.

The Pop artists minimized personal nuances and evidence of the artist's hand. Photographed, silk-screened, and stenciled images and flat applications of paint were generally employed to convey a passive and detached point of view. This impersonal approach was intended to reinforce the sense of anonymity and standardization encouraged by our mass-media culture. Thiebaud, by contrast, loves the manipulation of paint. It is a passion that stems from his early interest in painterly artists ranging from Sorolla to de Kooning.

Furthermore, while Pop artists were preoccupied with signs and sign systems, Thiebaud was, and remains, absorbed with objects per se, the visual perception of them, and the translation of this perception onto canvas. The iconoclasm of artists such as Warhol or Lichtenstein is foreign to Thiebaud who has always and openly admired those artists who have helped mold the tradition of realist painting. Among them he cites Velázquez, Vermeer, Chardin, Eakins, and Hopper.[24]

H.C. Westermann
Cliff. 1970
glass, wood, putty, chrome, oil
37 x 16⅛ x 16⅛ inches
Collection of Indiana University Art Museum

Wayne Thiebaud
Half Dome Evening. 1975
oil on canvas
28 x 16 inches
Private collection

Thiebaud's personal style of painting was developed between 1957 and 1960 during a time of personal critical evaluation and assimilation. After his experiences with Abstract Expressionist painters in New York, in 1956 and 1957, especially de Kooning and Kline, he concluded that representational painting, rather than their expressionist style, was for him a thought-provoking means. The realistic rendering of objects became the vehicle to explore problems of composition, space, light and paint handling.

An interest in the ambiguity of visual perception and the transference of information to representational work led to Thiebaud's studies of nineteenth-century Italian, Spanish and French schools of painting. The perception that a representational work of art can be both real and abstract through the handling of a medium is the essence of Thiebaud's work.

Thiebaud works directly from observation, making hundreds of preliminary sketches either on site or from memory. He adopts, adapts and changes his subjects by means of reduction and simplification. Selective perceptual nuances appear, enhancing the visual record.

Since the early 1960s Thiebaud has focused on a wide range of still life subjects, which make up the largest portion of his work. In 1963 he turned his attention to the human figure, breaking from consumer objects as subject matter. As in the still lifes, Thiebaud sees the figure set in light and space, avoiding expression of personality. The nonnarrative approaches continue in landscapes begun in 1966 after concentrated production of figurative work.

In western landscapes, Thiebaud works within a traditional emphasis on the physical qualities of a setting. He is interested in more than the pictorial aspects of a landscape, always manipulating it. Geometric and realistic reconstructions of the landscapes are translated by arcs, cubes, diagonal as well as rectangular lines, adding structural interest to the artist's sense of light and texture. His cityscapes are the result of simultaneous experiments with time and space perception and a move to San Francisco in the early 1970s. Next to the still lifes, this body of work is his largest and most varied. Urban landscapes build on the artist's earlier forms and concepts and the power of abstraction, but they also brought a dramatic dimension to his art of capturing the city's topographical extremes. Feelings of pressure and compaction create compositions of exaggerated vertigo.

In 1975 Thiebaud was among a number of artists who received commissions from the United States Department of the Interior to celebrate the country's bicentennial the following year. Artists selected their own landscape subjects from properties under the department's jurisdiction. Thiebaud chose California's Yosemite National Park; *Half Dome Evening* of 1975 and *Yosemite Rock Ridge* of 1975, 1977 and 1985 resulted from this project.

Wayne Thiebaud
Yosemite Rock Ridge. 1975, 77, 85
oil on canvas
36 x 36 inches
Private collection

Although many of these landscapes may appear as exaggerated figments of Thiebaud's imagination, they are the result of actual observations and subsequent visual recollections of natural wonders such as the Grand Canyon, the eroded hills of California's gold and wine country, and the granite heights of Yosemite National Park. The paintings are based on small drawings and paintings executed on site but completed in the artist's studio.[25]

In *Half Dome Evening* the image is tightly cropped to bring attention directly on the towering mass of the Yosemite granite formation. Cropping as a dramatic device, manipulation of size and elimination of a horizon line—techniques recalled from his days as a commercial artist—have had a tremendous impact on Thiebaud's landscapes, still lifes and figure paintings. As evidenced by the three *Yosemite Rock Ridge* dates, he occasionally reworks paintings. From the ground looking upward, the richly colored granite ridge looms large, engulfing all around it. Tiny trees hang in the crevices, reinforcing the exaggerated scale.

From landscape invention to the visual reality of his figure paintings, Thiebaud's work sets a special tension between Modernism and Traditionalism, a hallmark recognized as Thiebaud's alone.

EDWARD RUSCHA ■ Edward Ruscha's thirty-year body of work has been categorized as Pop, Conceptual and Postmodern. Although Pop is the first category to fit his painting style, Ruscha's use of words combined with images drawn from popular culture emphasizes the relationship between image and language. His preoccupation with clichés, riddles and visual and verbal puns link his work with the absurdity previously sought by the Dada artists. Ruscha's ambiguous juxtaposition between images and alphabet differs from Conceptual Art's rejection of the primacy of the object. As a Postmodernist who seeks appropriation as subject matter, his extraordinary resourcefulness in combining words and images is the single-most factor responsible for his importance as an artist.

Ruscha (born 1937) was trained in graphic design at Chouinard Art Institute in Los Angeles. His early work approached painting in favor of simplified forms reproduced in direct lettering and borrowed from advertising techniques. After seeing a reproduction of Jasper Johns' *Target With Four Faces* in 1957, Ruscha was inspired to work with images and print media himself. His was a more deliberate approach, however, as he worked from preconceived images. From Johns he learned that painting could be language, and from Warhol that painting could be photography.

By 1962 Ruscha had established parameters that would define his work. He sought the perfect union between images and words as he appropriated mass-media imagery and incorporated typography onto large canvasses. Words and images are abstractions used by Ruscha to communicate his understanding of reality in America, addressing the preoccupations of day to day existence. His work of the early 1960s responded to visual stimuli in Los Angeles—the logo for 20th Century Fox, the Standard Oil gas station and the Hollywood sign—that appeared on canvases in a variety of applications.

Gradually the painted elements became more minimalized and thought-provoking. Begun in the late 1960s a series of word paintings using liquids—egg yolk, carrot juice and blood in combination with oil paint and pastel—were rendered on satin, taffeta and other materials that suggestively combined his ideas and wit. He chose words such as "Desire," "Western" and "Murder" and painted the character of the word in such a way that it conveyed something about the myth of the West. Works on paper made with gunpowder and pastel followed, along with paintings of capital letters spaced across canvases of graduated color, conveying simple, humorous messages. Between 1962 and 1972 Ruscha published fifteen books of photographs with straightforward titles documenting his travels in the West and interest in imagery. Visual wordplay developed into conceptual and literary undertakings by the 1970s; graphics, drawings and paintings featured lines of meaning-laden phrases superimposed on image fields.

A series of horizontal word-and-image paintings of blended layers with lushly worked surfaces of saturated color emerged in 1977. Vast desert horizons at mid-afternoon, twilight, dusk or sunrise evoke uneasy expectation. In these paintings Ruscha began to rely more on relationships between illusionistic images and words, than on word images.

He embarked on a series of paintings in the mid-1980s that demonstrated a major departure from earlier creations. In silhouette with a soft-focus quality achieved by airbrushing, the images have a powdery effect. Sometimes they appear as objects photographed under very low light conditions, and unlike his earlier "cool" paintings, which resembled slick advertisements, these reveal an unsettling sense of grim melodrama. They are so enigmatic that Ruscha sometimes omits words completely, relying on the image alone to deliver his message. An empty house, a lone howling coyote, a pair of ships and desert landscapes deal with an American West.

In a sense, Ruscha launches his ironical attack on the American Dream by demonstrating how successful it has been. For the American Dream has indeed given the American world a dreamy appearance and put the American mind into a seemingly permanent dream state, in which it thinks and sees things as though in a dream. It has convinced people to regard reality as unreality—that's the real American way. Thus, Ruscha's pictures not only profoundly critique the American

Edward Ruscha
A Certain Trail. 1986
acrylic on canvas
59 x 145½ inches
Collection of
Dr. and Mrs. William C. Janss, Jr.
Photograph courtesy of
James Corcoran Gallery

Dream, but show it in action. They toy with our sense of reality, or rather show how absurd it already is—how infected it is by dream images and dream ideas, all amounting to nothing. Ruscha's art is ultimately about illusion, especially the dream of transcendence of reality that comes with the dream of eternal well-being and happiness.[26]

In *The Uncertain Trail* of 1986, a diagonal composition consistent with earlier works, Ruscha uses the sentimental image of a wagon train as an archetype of man confronting the westward expansion. In this painting achieving the American Dream is posed as figures struggling toward the horizon.

Clearly, Ruscha's art is motivated by a search for the unassuming icon which can nevertheless be transformed into an object of extended contemplation. Although in a compositional sense he is every bit as interested in an iconic description of the world as Warhol, Ruscha methodically de-mythifies his subjects by bringing them into myopic perspective and making them seem no more than found objects drifting unattached across the viewer's consciousness. By thus sidestepping the problem of reassigning values to his subject, Ruscha has successfully opened a realm of linguistic and painterly possibilities that will continue to exert a telling influence on many generations to come.[27]

Roger Brown
Saguaro's Revenge. 1983
oil on canvas
72 x 48 inches
Collection of Frederick R. Weisman
Art Foundation

Roger Brown
Garden of Eden. 1990
oil on canvas
48 x 48 inches
Courtesy of Phyllis Kind Gallery,
New York

ROGER BROWN ■ Chicago-based Roger Brown (born 1941), paints dramatic narratives and events drawn from television, billboards, comic strips, magazines, newspapers, his immediate environment and cross-country road trips. His stories of horrifying news events, biblical themes, political commentaries, historical events and art world machinations are set in simplified and stylized cities or landscapes.

As an observer recording the vernacular of his time, Brown places his subjects into highly charged and disconcerting environments. He combines ominous backlit skies, silhouetted stiff figures, shifting scales and isometric perspectives to create stage-like settings that encourage a sort of viewer voyeurism. The forms seem easy to understand yet there is an underlying complexity as the works explore preconditioned perceptions. Brown reveals a troubled vision of contemporary life; his startling emblematic figures attract the viewer with mysterious silhouettes. The two-dimensionality of these paintings is enhanced by back-lighting; spatial illusion is suggested by atmospheric perspective. "The repetition of forms within a stagelike setting and eerie light emanating from the clouds symbolize our entwined feelings of wonder and boredom. Clearly, the artist refuses to be didactic in these paintings."[28]

Since the late 1960s Brown's work has focused on altered, inhabited landscapes. They are an imaginative synthesis of memories of an actual location, stamped with his personal sense of place. The motifs are used to explore various ways that we experience a landscape. The viewer has been either guided or misguided by Brown.

Saguaro's Revenge of 1983 and *The Garden of Eden* of 1990 reveal Brown's distress at the lack of respect and care for the western environment and its life forms. The monumental saguaro cacti and bison, symbols of the West, are set in characteristic landscapes. The vast presence of the sharply patterned western skies suggests endless grandeur, boredom and mysteriously haunting and hypnotic forces. Brown questions the conflicting, often ambivalent ways that people have responded to the fragile land.

Joyce Treiman
The Nude Out West. 1987
oil on canvas
70 x 70 inches
Collection of Mr. and Mrs. John Pritzker
Photography by M. Lee Fatherree

JOYCE TREIMAN ■ Joyce Treiman (1922-1991) studied under Philip Guston who reinforced her respect for the tradition of figurative painting while he added a great deal to her technical repertoire.

Treiman began painting as a Social Realist in the 1940s, explored abstract paintings in the 1950s and in the early 1960s began to establish her unique style combining Expressionism and Surrealism. During the period of Abstract Expressionism, she was a nonconformist who undertook the Old Testament as her theme. When Pop Art emerged as a mainstream art, she moved toward a more figurative style.

Without regard to current styles in the art world, Treiman, in her idiosyncratic way, continually participated in a personal quest for an image of humanity in the context of historical and contemporary concerns. Her "cast of characters," which emerged in the early 1960s and appeared in paintings and drawings throughout her life, act in various episodes. The general milieu for these events is turn-of-the-century yet sufficiently ambiguous for events to appear dream-like and without reference to time and place. About Treiman's early work William Wilson states:

> Her extraordinary paintings often involve types familiar to a middle class intelligentsia such as Thomas Eakins used to paint in Philadelphia. ... Such a world turns on domestic ritual including respect for Renaissance Humanism as we learned it at university. Joyce Treiman's paintings are partly about an historical struggle to forge American culture out of European civilization. Every artist who is aware of ancestors who came here as foreigners shares a certain lonesome wistfulness. Some of Joyce Treiman's recent paintings involving American Indians are germane. Some are heartbreaking.[29]

In drawings and paintings from 1967 to 1990, Treiman's portraits of such master painters as Rembrandt, Turner, Goya, Monet, Bonnard, Degas, Toulouse-Lautrec, Sargent and Eakins often appear among images of her friends and herself. Although inspired by painters of the past, her draftsmanship displays a fine, linear quality of her own, which is combined with painterly use of color and tone in her pastel drawings. Clearly an extension of her painting, the diversity and freedom found in drawing are established in her style.

Treiman also was intrigued with the monotype and created an extensive body of work in the medium. Using the improvisational and sometimes simple nature of the monotype, she handled the inks with a humanistic spontaneity. The freedom inherent in her technique, the liberation from producing whole editions, the elimination of a press (Treiman used a kitchen rolling pin instead of a press to create the pressure), and the sensuous and beautifully modulated mass and color made monotype a preferred medium. Continuing her concern with the human figure and the human condition, Treiman achieved a loose rhythmic style while painting on a metal plate instead of a canvas. Many of her large monotypes describe and comment on a world of the past.

The Nude Out West of 1987 is one of a series of paintings in her mature work that made thematic use of the American West, with all its heroism, failures and myths. This work humorously expresses Treiman's view of the West, its heritage and subconscious place in the twentieth century. Treiman has combined several painting styles and has put her self-portrait with artist's palette in hand into the painting, playing out her childhood game of "cowboys and Indians." In discussing these works, Treiman said that "artistically, this series of paintings was also a personal protest against the current trends of the art market, the 'less is more' syndrome and the banishment of figuration as a viable means of expression."[30]

Treiman's art is a compelling and versatile vision of autobiographical themes and observations on the human condition. In describing Treiman's courage to follow her individual instincts, Selma Holo states:

> Joyce Treiman, then, by means of art *and* life, reminds us along with Yeats that soul can "clap its hands and sing for every tatter in its mortal dress." She challenged us to remember, long before it was fashionable to do so, that rich and lasting originality stems from the carefully cultivated understanding of where and when to jump off the train of our collective memory into uncharted terrain.[31]

PETER DEAN ■ Peter Dean (born 1934), has remained on the outskirts of the art world's mainstream most of his life. He has leaned on his personal vision to express autobiographical, political and social concerns that connect past with present. Consistently and concurrently exploring, Dean focuses on life's evils, making connections to politics, the American Nazi movement and dealing with violence and alienation of contemporary urban life. A more tender sensibility focuses on the magical moments associated with fantasy, dreams and imagination. Working expressively and figuratively, Dean has much to say about a variety of subjects. Since the 1960s he has created a number of works in groups entitled "Personal Event Paintings," "Political Paintings," "Small and Large Murders," "Assassination Paintings," "History Paintings," "Landscapes" and "Native Americans."

> Critics and historians define Expressionism as an art that looks inward, to sensibility's private chambers, a definition that can't wrap itself around Dean's love of spectacle. In the light shed by familiar nomenclature, he cuts the paradoxical figure of an extroverted Expressionist.[32]

Dean became visible in New York in the mid-1960s when he was a co-founder of the "Tonque" group including expressionistic painters Joseph Kurhajec, Peter Saul and Leon Golub. They made political paintings at a time when topics of death, threat of death or violence were considered inappropriate choices of subject matter for artists.

In 1969 Dean and six other painters founded a group called the "Rhino Horn," a socially critical expressionist group. Sexuality was their major topic. Although their iconoclastic works were not successful in penetrating the Minimal/Conceptual strongholds of the time, they were recognized as rallying statements against the Vietnam War. By 1975 "Rhino Horn" had dissolved after exhibitions had been presented in New York, New England, the Midwest, the South and California.

Toward the end of the 1970s, Dean's exuberant and raucous style of painting received more recognition when Neo-Expressionism emerged as the 1980s style. He returned to political subjects after several years of working primarily from his imagination, memories and dreams.

> The chasms in Dean's imagery also read as gaps between the self and a difficult world. His art is not absorptive but confrontational. At their most seismically active, the fault lines running through his paintings displace aspects of Dean himself to such a distance that he becomes alien in his own eyes—another presence to confront. His most violent paintings look like, among other things, tough negotiations between Dean the passive consumer of bad news and Dean the agent of a restlessly active conscience. His moral concerns provide another reason to call him an extroverted Expressionist, a painterly painter with an immediately recognizable style but no inclination to treat its display as the purpose of his art. Nor does he show much interest in his heroes' styles.[33]

From his Native American series, the painting *Battle Won, War Lost* of 1990 reflects the artist's empathy with Native Americans. The frontal figure of an Indian chief warrior anchors the difficult, chaotic composition and confronts the viewer with both triumphs and pains. The chief stands as a magnificent symbol of pride and defiance in resistance to repressive authority. The Indian persona, a creation of Dean's, is symbolic of life-renewing forces and suggests these forces will prevail. Native subjects have absorbed Dean since he first became aware of the plight of Native Americans while working as a field geologist in Montana and North Dakota. He says he was drawn to the "tragedy of being the victor but losing the whole thing." Dean's characters act in a dramatic tableau commemorating the moment when his imagination merges with myth. They "inhabit that time and place where fantasy is as real as so-called reality, when things are reversed, as in ritual or madness."[34]

ALEXIS SMITH ■ Alexis Smith (born 1949), has long appropriated literary narratives and visual images from popular culture and assembled them in ways that dislodge their expected meanings. Hunter Drohojowska, analyzing Smith's narrative collages from the early 1970s in relationship to Conceptual art, wrote that—

> Smith's work does have a place in Conceptual art as it is more broadly understood today; in fact, it has a kind of prescience for the current generation of Conceptual artists, who are sometimes referred to as post-Conceptualists. In this, it is related to the late-'60s and early-'70s approach of John Baldessari and Edward Ruscha, who, of course, also live in California. The apprehension of society as mediated by images seems to have been quite evident, as an urgent subject for art, to artists in the Los Angeles of the post-Pop '70s. The city was nearly naked of museums and galleries; with few at-hand sources for so-called "high culture," these artists addressed the available popular culture of films, advertising, paperbacks, magazines, and billboards.[35]

Smith's ironic narrative art spins fantasies from popular American culture and usually incorporates found or invented subtexts, quotes or myths. Her private and recurring concern with memory, destiny and the passage of time affects her choice of imagery. Works often deal with the vernacular of the 1940s and 1950s through appropriated images and objects from magazines, advertising, films and literature. Among her themes are Raymond Chandler mysteries, famous women named Jane, the opera "Madama Butterfly" and Jack Kerouac's novel *On the Road*. Her witty, mixed-media constructions seek poignant experiences in the most mundane aspects of American culture. Sifted visual and verbal clichés are brought together and reshuffled and recombined, and for the past twenty-five years, her messages have been relayed through poetic metaphors.

In 1985 Smith produced a series of thirty collages on stereotypes of famous women, either factual or fictional, named Jane, among them Jane Russell, Jayne Mansfield, Calamity Jane, Tarzan and Jane, Dick and Jane and Jane Doe. From memories and myths surrounding Jane personalities, Smith has composed these images from the popular culture of her youth with quotes from both classic literature and trashy novels. The sentences or phrases she chooses create unprecedented juxtapositions, enhance mundane images and give her clever collages a life of their own.

Cinderella Story of 1985 depicts actress Frances Farmer as Calamity Jane in the film *Badlands of Dakota* of 1941. This Hollywood-inspired Jane, dressed in a fringed leather jacket and cowboy hat with red eyelashes and her pistol drawn, is accompanied by "It was well past midnight and she was very tired" silk-screened in pink across the lower

Alexis Smith
Cinderella Story. 1985
mixed media collage
32½ x 21 inches
Collection of James and Linda Burrows,
Los Angeles
Photography by
Douglas M. Parker Studio,
courtesy of Margo Leavin Gallery,
Los Angeles

Alexis Smith
Desert Rose. 1991
mixed media
16¾ x 15½ x 2 inches
Collection of James and Linda Burrows,
Los Angeles
Photography by
Douglas M. Parker Studio,
courtesy of Margo Leavin Gallery,
Los Angeles

edge. With other bits of imagery, including a gold corsage, pearl earring, and Princess Grace stamp and emblem, the title *Cinderella Story* releases the interpretation of glamorous figures in real life and in fantasy.

Desert Rose of 1991 is one of a dozen works of nature-oriented subjects. Smith has said that she's always been keen on the desert and draws upon her experience when she was growing up and spending time at her parents' second home in Palm Springs, California. The piece reveals her personal theory that the imagery of something should be appropriate to the content of the place. The witch reflects the artist's humor or bitchiness as she flies past darkly silhouetted saguaro cacti in an orange sky and a 1950s kitsch-derived silk rose that is softly toned in orange and black. While the scene contains no text, the intimate collage incorporates telling clues scavenged from the West.

Sometimes coy, often ironic, Smith's assemblages pilfer mementos from history, mythology or metaphors of American culture. After exploring American culture for the past twenty-five years, she is presently drawn to more universal cultural themes.

DOTTY ATTIE ■ Throughout history artists have adapted paintings of other artists to their own use. Previous generations of artists borrowing from old masters eschewed all signs of individual expression. However, contemporary artists are eager to make the viewer aware of the origin and the deliberate recreating of it.

For the past two decades, Dotty Attie (born 1938), has borrowed, copied or extracted scenes from old master and historical paintings by Ingres, Eakins, Copley, Caravaggio, Wimar, Vermeer and others. She recombines with painstaking selectivity certain details, fragments or scenes before adding fictional texts on multipaneled canvases. Her unexpected rearrangement and dramatic juxtaposition of discrete copied images from the past leads to highly original and mysteriously charged narratives. Her images are jolting with free-for-all historical irony and cynicism, vaguely recalling their origins.

> For her, these details cannot be photographed or collaged from reproductions. Nor can they be merely approximated or abbreviated in paint. Any of those shortcut methods would signify dissociation, whereas Attie wants to transmit a sense of real intimacy of contact with her repossessed images—and more than that, her investment in them. This end can be realized only in the exquisite mimicry of the old masters' suaveness of facture and evenness of hand. Her own hand, therefore, goes through the same idiosyncratic paces as theirs did. As a result, Attie knows her allusions from the inside and has "earned" them with a limpid effort that melts into the brushed, lifelike sheen of her painted surfaces.[36]

The series of six-inch-square canvases are meticulously rendered fragments overlaid or interspersed with texts to capture a psychological narrative. Her text interrupts the interpretation of the painting with messages and authoritatively worded quotations that stimulate the viewer to complete the story.

The aim of Attie's art is to elicit excitement at the point of perception that a new story is unfolding behind the original story. Because of the restyled fragments and gaps in her text, however, the viewer is not permitted to know the complete message with which she seduces and teases. "The conceptual framework of interpolation and juxtaposition constantly brings into conjunction horror and aspiration, qualities which characterize the entire body of work. Attie has said that 'from base impulses great work can come,' a truth as relevant to her own work as it is to that of the masters she makes her sources."[37]

They Traveled West of 1990-91 and composed of six sections, recapitulates a painting entitled *The Attack of an Emigrant Train* of 1856 by Carl Wimar. The text which has been painted over the dark lower two panels, reads, "They traveled West to live a better, more independent life. The opportunity to claim large tracts of land that now belonged

to no one, was not to be missed." The clarity of Attie's rendering and the matter-of-fact text suggest a straightforward message with a predictable conclusion, but the scene does not allow a viewer's complete interpretation. As they attack a covered wagon, the Indians menace the mid-nineteenth-century travelers. Inside the wagon the wounded are tended while the able-bodied fire at their assailants. The fragment Attie pulled from Wimar's painting does not show three or more wagons *behind* this wagon, showing the intrepid settlers holding their own. The conflict between the Indians and the pioneers hints at a terror incompatible with the simple, declarative sentence. The viewer is provoked to ponder as Attie challenges the role paintings such as Wimar's have played in forming American myths of nationalism.

> Attie offers us twice told tales that, in the retelling, show the arbitrariness of the signs used for the retelling and our uncertainty of the original story. Attie makes clear to us the legendary character of knowing, whose substance dissipates in to wild rumors. Attie's narratives are a whispering campaign against the truth. But as the whispers accumulate, we seem nearer to it. Attie shows us that we end up believing our own lies.[38]

Believing our own lies, our textbook truths, perhaps best describes our interpretation of the West and its documentation. Contemporary artists selected for this exhibition have looked to the American West not only for historical interpretation but for fiction that serves as a source for replication. In so-doing they have given us an altered sense of mythology and history and how they continued to be transformed through the passage of time.

Dotty Attie
They Traveled West. 1990-91
oil on linen
24 x 13 inches
Courtesy of P.P.O.W., New York

105

[1]Lawrence Alloway, "Pop Art, the Words," *Topics in American Art Since 1945* (New York: W.W. Norton and Company, Inc., 1975), p. 120.

[2]Ernst Busche, *Roy Lichtenstein: Das Fruehwerk 1942-1960* (Berlin: Gebr. Mann, 1988) p. 144.

[3]Lawrence Alloway, *American Pop Art* (London: Collier MacMillan Publishing Co., Inc., 1974), p. 80.

[4]Jack Cowart, *Roy Lichtenstein 1970-1980* (New York: Hudson Hills Press, Inc., 1981), p. 127.

[5]Ibid.

[6]Ibid., p. 128.

[7]Carter Ratcliff, "The Work of Roy Lichtenstein in the Age of Walter Benjamin's and Jean Baudrillard's Popularity," *Art in America,* February 1989, pp. 116-117.

[8]David Robbins, "Cigar-Store Indian: Andy Warhol's Cowboys and Indians," *Art*, May 1986, p. 53.

[9]Carter Ratcliff, "Red Grooms' Human Comedy," *Portfolio*, March/April 1981, p. 61.

[10]Ibid. pp. 120,121.

[11]Peter Plagens, *Sunshine Muse: Contemporary Art on the West Coast* (New York and Washington: Praeger Publishers, 1974), p. 140.

[12]Colin Naylor, ed., *Contemporary Artists* (Chicago and London: St. James Press, 1989), p. 323.

[13]Anne Ayres, "Llyn Foulkes," *L A Pop in the Sixties* (Newport Beach: Newport Harbor Art Museum, 1989), pp. 97, 100.

[14]Suzanne Muchnic, "Foulkes' 'Portraits' of Contemporary Terror," *Los Angeles Times*, February 26, 1986, p. 8.

[15]Llyn Foulkes, *American Myths* (New York: Kent Fine Art Inc., 1986), p. 8.

[16]William Wilson, "The Galleries: La Cienega Area," *Los Angeles Times*, March 25, 1983, p. 15.

[17]Marcia Tucker, *William Allan* (New York: Whitney Museum of American Art, 1974), n.p.

[18]Ibid.

[19]Robert C. Hobbs, "Robert Arneson: Critic of Vanguard Extremism," *Arts*, Nov. 1987, p. 93.

[20]Ibid.

[21]Donald B. Kuspit, "Arneson's Outrage," *Art in America*, May 1985, p. 136.

[22]Barbara Haskell, *H.C. Westermann* (New York: Whitney Museum of American Art, 1978), p. 9.

[23]Bill Barrette, *Letters from H.C. Westermann* (New York: Timken Publishers, Inc., 1988), p. 13.

[24]Karen Tsujomoto, *Wayne Thiebaud* (Seattle and London: University of Washington Press, 1985), p. 39.

[25]Ibid., p. 124.

[26]Donald B. Kuspit, "Signs in Suspense: Ed Ruscha's Liquidation of Meaning," *Arts*, April 1991, p. 57.

[27]Dan Cameron, "Love in Ruins," *Edward Ruscha* (Rotterdam: Museum Boymans-Van Beuningen, 1990), p. 17.

[28]John Yau, "Roger Brown and the Spectacle," *Roger Brown* (New York: George Braziller, Inc., 1987), p. 22.

[29]William Wilson, *Joyce Treiman Paintings 1961-1972* (La Jolla, California: La Jolla Museum of Contemporary Art, 1972) p. 4.

[30]Joyce Treiman, conversation with author, Palm Springs, California, April 8, 1991.

[31]Selma Holo, *Joyce Treiman: Friends and Strangers* (Los Angeles: University of Southern California, 1988), p. 2.

[32]Carter Ratcliff, "The Art of Peter Dean," *Peter Dean* (Grand Forks: North Dakota Museum of Art, 1989), p. 9.

[33]Ibid., p. 14.

[34]Lucy Lippard, "Builder on the Edge of a Chaotic World," *Peter Dean A Retrospective* (New York: Alternative Museum, 1990), p. 12.

[35]Hunter Drohojowska, "Alexis Smith R Tist," *Artforum*, October 1987, pp. 86-90.

[36]Max Kozloff, "The Discreet Voyeur," *Art In America*, July 1991, p. 102.

[37]Ellen Handy, "Reading Between the Lines" *Dotty Attie: Painting and Drawings* (Pittsburgh: Pittsburgh Center for the Arts, 1989), p. 12,14.

[38]Donald B. Kuspit, "Review of Exhibitions: Dotty Attie at AIR," *Art in America*, February 1984, p. 144.

Catalog of the Exhibition

WILLIAM ALLAN
Shadow Repair for the Western Man.
1970
oil on canvas
90 x 114 inches
Collection of University Art Museum,
University of California at Berkeley

■

GUY ANDERSON
Deception Pass through Indian Country.
1959
oil on paper on plywood
11 x 30⅜ inches
Collection of Seattle Art Museum, gift
of the Sidney and Anne Gerber
Collection

■

ROBERT ARNESON
Last of the Great White Buffalo Hunters.
1987
wood, fixall, ceramic, paint, leather
57 x 74 x 32 inches
Collection of Denver Art Museum,
funds from Colorado Contemporary
Collectors and National Endowment for
the Arts Museum Purchase Program

■

DOTTY ATTIE
They Traveled West. 1990-91
oil on linen
24 x 13 inches
Courtesy of P.P.O.W., New York

■

WILL BARNET
Self Portrait. 1948-49
oil on canvas
45 x 38 inches
Courtesy of the artist and
Terry Dintenfass Gallery, New York

■

ROGER BROWN
Garden of Eden. 1990
oil on canvas
48 x 48 inches
Courtesy of Phyllis Kind Gallery,
New York

ROGER BROWN
Saguaro's Revenge. 1983
oil on canvas
72 x 48 inches
Collection of Frederick R. Weisman
Art Foundation

■

BYRON BROWNE
Variations on Haida Masks. 1934
watercolor and india ink on paper
19¾ x 11 ¾ inches
Courtesy of Michael Rosenfeld Gallery,
New York

■

CONRAD BUFF
Jawbone Canyon. ca. 1925
oil on canvas
47 x 66 inches
The Buck Collection, Laguna Hills,
California

■

PAUL BURLIN
Grand Canyon. n.d.
oil on canvas
20 x 25 inches
The Harmsen Collection

■

KENNETH CALLAHAN
Trail Crew. 1940
oil on canvas
20 x 22⅞ inches
Collection of Seattle Art Museum,
Eugene Fuller Memorial Collection

■

ELLIOTT DAINGERFIELD
*A Vision of the Dawn (or Nude in Grand
Canyon).* 1913
oil on canvas
48 x 36½ inches
Collection of James E. Lewis Museum of
Art, Morgan State University

■

ARTHUR B. DAVIES
The Edge of the Redwood Forest. 1905
oil on canvas
18 x 30 inches
The Regis Collection, Minneapolis

ARTHUR B. DAVIES
Indian Fantasy. ca. 1918
oil on canvas
17⅝ x 16⅛ inches
Collection of The Newark Museum,
bequest of Miss Lillie Bliss, 1931

■

LEW DAVIS
Little Boy Lives in a Copper Camp.
1939
oil on masonite
29½ x 24½ inches
Collection of Phoenix Art Museum, gift
of IBM Corporation

■

STUART DAVIS
New Mexico Gate. 1923
oil on linen
22 x 32 inches
Collection of Roswell Museum
and Art Center, gift of
Mr. and Mrs. Donald Winston and
Mr. and Mrs. Samuel H. Marshall

■

PETER DEAN
Battle Won, War Lost. 1990
oil on canvas
80 x 60 inches
Courtesy of Peter Dean

■

RICHARD DIEBENKORN
Albuquerque. 1951
oil on canvas
40½ x 50¼ inches
Collection of Oklahoma City Art
Museum

■

RICHARD DIEBENKORN
Albuquerque. 1951
oil on canvas
38½ x 56¼ inches
Private collection

■

PHIL DIKE
Copper. 1935-36
oil on canvas
38 x 46¼ inches
Collection of Phoenix Art Museum,
purchase with funds provided by
Western Art Associates

MAYNARD DIXON
The Ancient. 1915
oil on canvas
19¼ x 11½ inches
Collection of Katherine H. Haley

■

MAYNARD DIXON
Cloud World. 1925
oil on canvas
34 x 62 inches
Collection of Arizona West Galleries,
Scottsdale

■

ARTHUR DOW
Bright Angel Canyon. 1912
oil on canvas
30 x 40¼ inches
Collection of Ipswich Historical Society

■

LLYN FOULKES
The Last Outpost. 1983
mixed media assemblage
81 x 108 x 5 inches
Collection of
Palm Springs Desert Museum,
purchase with funds provided by
the Contemporary Art Council, 1989

■

LLYN FOULKES
The Page. 1963
oil on canvas
87½ x 84 inches
Collection of The Oakland Museum,
gift of Anonymous Donor

■

ADOLPH GOTTLIEB
Symbols and the Desert. 1938
oil on canvas
39¾ x 37⅞ inches
Collection of Adolph and Esther
Gottlieb Foundation, Inc.

■

ADOLPH GOTTLIEB
*Untitled (Still Life with Watermelon —
Dry Cactus).* 1938
oil on canvas
23⅞ x 30 inches
Collection of Adolph and Esther
Gottlieb Foundation, Inc.

MORRIS GRAVES
Joyous Young Pine. n.d.
watercolor on paper pasted on
cardboard
52¾ x 27 inches
Collection of Santa Barbara Museum of
Art, gift of Wright S. Ludington

■

RED GROOMS
Great Western Act. 1971
oil on masonite
23½ x 35 inches
Collection of Red Grooms, New York

■

RED GROOMS
Shoot-out. 1980
painted bronze
11½ x 22 x 9¼ inches
Private collection

■

MARSDEN HARTLEY
New Mexico Recollections #6. 1922
oil on canvas
26 x 36½ inches
The Harmsen Collection

■

MARSDEN HARTLEY
Window, New Mexico. 1919
oil on canvas
37⅛ x 29 inches
Collection of Hirshhorn Museum and
Sculpture Garden, Smithsonian
Institution, gift of
Joseph H. Hirshhorn, 1966

■

RAYMOND JONSON
Light. 1917
oil on canvas
45 x 42 inches
Collection of Museum of Fine Arts,
Museum of New Mexico, gift of
John Curtis Underwood, 1925

■

RAYMOND JONSON
Pueblo, Acoma. 1927
oil on canvas
37 x 44 inches
The Harmsen Collection

RAYMOND JONSON
Southwest Arrangement. 1933
oil on canvas
45 x 20 inches
Collection of Jonson Gallery of the
University Art Museum, University of
New Mexico, Albuquerque

■

WALT KUHN
Commissioners. 1918
oil on canvas
10⅛ x 16 inches
Collection of
Colorado Springs Fine Arts Center,
gift of Vera and Brenda Kuhn

■

ROY LICHTENSTEIN
American Indian Theme III. 1980
woodcut
35 x 27 inches
Collection of
Palm Springs Desert Museum,
purchase with funds provided by
the Walter N. Marks Graphics Fund

■

ROY LICHTENSTEIN
American Indian Theme VI. 1980
woodcut
37½ x 50¼ inches
Collection of Walker Art Center,
Minneapolis, Tyler Graphics Archive

■

ROY LICHTENSTEIN
Cowboy on Bronco. 1953
oil on canvas
32 x 26 inches
Collection of The Butler Institute of
American Art, Youngstown, Ohio

■

ROY LICHTENSTEIN
The Straight Shooter. ca. 1956
oil on canvas
16 x 13½ inches
Collection of The Butler Institute of
American Art, Youngstown, Ohio

■

ROY LICHTENSTEIN
Two Sioux. ca. 1952-1954
oil on canvas
30 x 22 inches
Private collection

JOHN MARIN
Near Taos #5.
watercolor
13¼ x 17½ inches
The Harmsen Collection

■

JOHN MARIN
New Mexico, Near Taos. 1929
watercolor
14 1/16 x 21 1/16 inches
Collection of Los Angeles County
Museum of Art, The Mira T. Hershey
Memorial Collection

■

JOHN MARIN
Taos Indian Rabbit Hunt. 1929
watercolor
21 x 16 inches
Collection of University of Maine at
Machias Art Galleries, Machias, Maine

■

GEORGE L. K. MORRIS
Indians Fighting #1. 1934
oil on canvas
26⅛ x 22 inches
Collection of
Mr. and Mrs. Alan J. Pomerantz

■

GEORGE L.K. MORRIS
Indians Hunting #4. 1935
oil on canvas
35 x 40 inches
Collection of University of New Mexico
Art Museum, purchase through a grant
from the National Endowment for the
Arts with matching funds from the
Friends of Art

■

GEORGIA O'KEEFFE
Katchina. 1936
oil on canvas
7 x 7 inches
Collection of San Francisco Museum of
Modern Art, gift of the Hamilton-Wells
Collection

GEORGIA O'KEEFFE
Red and Black. 1916
watercolor
12 x 9 inches
Courtesy of The Gerald Peters Gallery,
Santa Fe, New Mexico

■

GEORGIA O'KEEFFE
The Red Hills Beyond Abiquiu. 1930
oil on canvas
30 x 36 inches
Collection of the Eiteljorg Museum of
American Indian and Western Art, gift
of Harrison Eiteljorg

■

AMBROSE PATTERSON
Rocky Landscape. 1946
oil on panel
22⅜ x 30 inches
Collection of Seattle Art Museum,
Eugene Fuller Memorial Collection

■

JACKSON POLLOCK
Camp With Oil Rig. 1930-33
oil on board
18 x 25¼ inches
Collection of
Mr. and Mrs. John W. Mecom, Jr.,
Houston

■

JACKSON POLLOCK
Composition with Horse. ca. 1934-38
oil on panel
10½ x 20¾ inches
Courtesy of The Gerald Peters Gallery,
Santa Fe, New Mexico

■

JACKSON POLLOCK
Untitled. ca. 1939-42
india ink on paper
18 x 13⅞ inches
Collection of Whitney Museum of
American Art, New York, purchase with
funds from the Julia B. Engel Purchase
Fund and the Drawing Committee

■

JACKSON POLLOCK
Untitled. 1943
ink and watercolor on paper
26 x 20½ inches
Collection of Montana Historical
Society, Poindexter Collection

RICHARD POUSETTE-DART
Primordial Moment. 1939
oil on linen
36 x 48 inches
Collection of
Mrs. Richard Pousette-Dart
■

RICHARD POUSETTE-DART
Untitled, Birds and Fish. 1939
oil on linen
36¼ x 60 inches
Collection of
Mrs. Richard Pousette-Dart
■

EDWARD RUSCHA
A Certain Trail. 1986
acrylic on canvas
59 x 145½ inches
Collection of
Dr. and Mrs. William C. Janss, Jr.
■

ALEXIS SMITH
Cinderella Story. 1985
mixed media collage
32½ x 21 inches
Collection of James and Linda Burrows,
Los Angeles
■

ALEXIS SMITH
Desert Rose. 1991
mixed media
16¾ x 15½ x 2 inches
Collection of James and Linda Burrows,
Los Angeles
■

WAYNE THIEBAUD
Half Dome Evening. 1975
oil on canvas
28 x 16 inches
Private collection
■

WAYNE THIEBAUD
Yosemite Rock Ridge. 1975, 77, 85
oil on canvas
36 x 36 inches
Private collection

MARK TOBEY
Mexican Ritual. 1931
oil on board
15¼ x 11¼ inches
Collection of Helen and Marshall Hatch
■

MARK TOBEY
Western Town. 1944
tempera
12 x 18¾ inches
Collection of Portland Art Museum,
bequest of Edith L. Feldenheimer
■

JOYCE TREIMAN
The Nude Out West. 1987
oil on canvas
70 x 70 inches
Collection of Mr. and Mrs. John Pritzker
■

JAY VAN EVEREN
Amerindian Theme. n.d.
oil and lacquer on masonite
46¼ x 45¾ inches
Collection of The Montclair Art
Museum, gift of Mr. and
Mrs. Rick Mielke
■

JAY VAN EVEREN
Untitled. n.d.
oil and lacquer on masonite
26¼ x 37¼ inches
Collection of The Montclair Art
Museum, gift of Mr. and
Mrs. Rick Mielke
■

ANDY WARHOL
Annie Oakley (#378). 1986
serigraph
36 x 36 inches
Courtesy of Kent M. Klineman
■

ANDY WARHOL
John Wayne (#377). 1986
serigraph
36 x 36 inches
Courtesy of Kent M. Klineman

ANDY WARHOL
Mother and Child (#383). 1986
serigraph
36 x 36 inches
Courtesy of Kent M. Klineman
■

ANDY WARHOL
Portrait of Dennis Hopper. 1970
acrylic on canvas
40¼ x 39⅞ inches
Collection of an anonymous lady
■

H.C. WESTERMANN
The Stranger. 1978
watercolor on paper
22¹/₁₆ x 30¾ inches
Collection of Seattle Art Museum, gift
of the Contemporary Art Council

Index of the Artists

William Allan, **86**, 87
Guy Anderson, 63, **64**
Robert Arneson, 87, **88**, 89
Dotty Attie, 104, **105**
Will Barnet, 57, **57**
Roger Brown, 37, **96**, 97, **97**
Byron Browne, 54, **54**
Conrad Buff, 31, **31**
Paul Burlin, 16-17, **17**
Kenneth Callahan, 58, **59**, 63
Elliott Daingerfield, 12, **13**, 14, 16, 23
Arthur B. Davies, **10**, 11-12, 14, 20, **21**
Lew Davis, 44-45, **45**, 47
Stuart Davis, 26-27, **27**, 29, 33
Peter Dean, 99, **100**, 101
Richard Diebenkorn, 65, **66**, **67**
Phil Dike, 43-44, **43**, 47
Maynard Dixon, **18**, 19, 31, 32, **32**
Arthur Dow, 14, **15**, 16, 23, 25
Llyn Foulkes, 82, **83**, **84**, 85, 87
Adolph Gottlieb, **46**, 47, **48**
Morris Graves, 58, **61**, 63
Red Grooms, 79, **80**, 81, **81**
Marsden Hartley, 19-20, **20**, 26, **28**, 29, 34, 58
Raymond Jonson, 23, **24**, 29-31, **30**, 34, 42, **42**, 65
Walt Kuhn, 22, **22**
Roy Lichtenstein, **70**, 71, **72**, **73**, **74**, **75**, 76, 91
John Marin, 33-34, **33**, 35
George L. K. Morris, 38, 40, **40**, **41**, 42, 54
Georgia O'Keeffe, 23, 25, **25**, 26, 34, 36-38, **36**, **37**, **38**, 47
Ambrose Patterson, 58, **60**
Jackson Pollock, 47, 49-50, **49**, **51**, **52**, **53**, 88, 89
Richard Pousette-Dart, 47, 54, **55**, **56**, 57
Edward Ruscha, 94-95, **95**
Alexis Smith, 101, **102**, 103, **103**
Wayne Thiebaud, 91-92, **92**, **93**, 94
Mark Tobey, 58, **62**, 63, **64**
Joyce Treiman, **98**, 99
Jay Van Everen, 38, **39**
Andy Warhol, 71, 76, **77**, **78**, 79, **79**, 91, 94, 95
H.C. Westermann, 89, **90**, 91, **91**

Boldface numerals indicate illustration

Palm Springs Desert Museum Board of Trustees

111